PREFACE

This book introduces the art of concurrent programming. This particular type of programming, with several activities progressing in parallel, is intellectually intriguing and is essential in the design of operating systems. Unix is used as a case study for exploring operating system structures. The Tunis implementation of Unix's nucleus (kernel) is presented as an example of a large concurrent program. Although the emphasis is on operating systems, the design and implementation techniques presented apply as well to other high performance, highly reliable software such as that in computer networks, real time control and embedded microprocessor systems.

The first two chapters overview concurrent programming and operating systems. Chapters 3 and 4 introduce the Concurrent Euclid (CE) language. Chapter 5 presents standard concurrency problems and their solutions. Chapters 6, 7 and 8 concentrate on Unix. Chapter 9 gives the structure of Tunis, a Unix-compatible nucleus written in CE. The last chapter shows how to construct a small kernel to support concurrent processes. An appendix gives the detailed specification of the Concurrent Euclid language.

The required background of the reader is a familiarity with a high-level language such as Pascal or Fortran as well as some familiarity with computer architecture. The programs presented in this book are written in Concurrent Euclid. This is a language that is suited for developing high performance system software as well as for teaching. Student's CE programs that use parallel processes are conveniently executable under systems such as Unix/11 and Unix/VAX. Alternatively, these programs can be down-loaded and executed on microprocessors such as the MC68000 and MC6809. The CE compiler is available from the CE Distribution

Manager, Computer Systems Research Group, University of Toronto, Toronto, M5S 1A4, Canada.

This book can serve as the main or subsidiary text for a course on operating systems or systems programming. Alternately, it may be used as a text book in a specialized course such as one on concurrent programming.

Acknowledgements. This book has evolved from an earlier book "Structured Concurrent Programming with Operating Systems Applications", which I co-authored with G.S. Graham, E.D. Lazowska and M.A. Scott. The present book has been made possible due to their essential contributions to its predecessor. I want to thank the people who have taken the trouble to suggest improvements in the earlier book; in particular the detailed comments by S.S. Toscani have been helpful to me in preparing this new book. J.C. Weber, S.G. Perelgut, D.R. Galloway, M.P. Mendell and D.T. Barnard have helped by reading drafts of the new book.

The Concurrent Euclid language was designed by J.R. Cordy and myself. This language is based on the Euclid language designed by B.W. Lampson, J.J. Horning, R.L. London, J.G. Mitchell and G.J. Popek with assistance from J.V. Guttag. B.A. Spinney, C.R. Lewis, B.W. Thomson, and C.D. McCrosky contributed to the CE language design and/or its compiler. D.B. Wortman, D.R. Crowe and I.H. Griggs helped inspire CE's design by their roles as implementors (with J.R. Cordy and myself) of the Toronto Euclid compiler.

P. Cardozo, M.P. Mendell, I.J. Davis and G.L. Dudek have done MSc projects involving Tunis design and implementation. S.W.K. Tjiang, D.R. Galloway and D.T. Barnard have also contributed to the Tunis work. The continuing interest of P.I.P. Boulton and E.S. Lee in the CE and Tunis work has been important to its progress. The following have been students in my graduate course in which we studied and evolved the Tunis design: P. Cardozo, A. Curley, R.S. Gornitsky, J.S. Hogg, S.A. Ho-Tai, P.M. McKenzie, J.L. More, B.A. Spinney, B.W. Down, G.L. Dudek, D.R. Ings, P. Kates, P.A. Matthews, M.P. Mendell, L.M. Merrill, R. Parker, B.R.J. Walstra, H.E. Briscoe, D. Chan, L. DeMaine, E.L. Fiume, R.D. Hill, L.Z. Zhou, S.G. Perelgut, and Y.C.L. Wong.

The terms VAX and PDP-11 are trademarks of the Digital Equipment Corporation. Unix is a trademark of Bell Laboratories.

The information on Unix in this book is based upon widely available materials, particularly upon excellent articles by the authors of Unix (D.M. Ritchie and K. Thompson). As they have stated, "The success of Unix lies not so much in new inventions but rather in the full exploitation of a carefully selected set of fertile ideas..."

CONCURRENT EUCLID, THE UNIX*SYSTEM, AND TUNIS

R.C. Holt

Computer Systems Research Group
University of Toronto

ADDISON-WESLEY PUBLISHING COMPANY
Reading, Massachusetts | Menlo Park, California
London | Amsterdam | Don Mills, Ontario | Sydney

This book is in the Addison-Wesley Series in Computer Science

Consulting Editor
Michael A. Harrison

Holt, R. C. (Richard C.), 1941–
    Concurrent Euclid, UNIX, and TUNIS.

    Bibliography: p.
    1. Concurrent Euclid (Computer program language)
2. UNIX (Computer system) 3. TUNIS (Computer program)
I. Title
QA76.73.C64H64   1983   001.64'2   82-13742
ISBN 0-201-10694-9

I.S. Weber has prepared the book for publication using a computer text editor and phototypesetter.

The research leading to CE and Tunis would not have been possible without the financial support of the Canadian Natural Sciences and Engineering Research Council and of Bell Northern Research Ltd.

R.C. Holt
May 1982
Toronto

CONTENTS

Chapter 1

CONCURRENT PROGRAMMING AND OPERATING SYSTEMS

Concurrent programming means writing programs that have several parts in execution at a given time. The concept of concurrent or parallel execution is intellectually intriguing and is essential in the design of computer operating systems. This book covers the fundamentals of concurrent programming using structured techniques. After an introduction to the need for concurrent programming and its basic concepts, a notation called monitors is presented and used for solving problems involving asynchronous program interactions. The concurrent algorithms in the book are presented in the Concurrent Euclid (CE) programming language. It is a language designed to support the development of highly reliable, high performance systems programs.

After giving examples of concurrency, this chapter concentrates on operating systems. Operating systems implement concurrent programs by sharing CPU time among several programs and use concurrent programs to control resources and serve users.

EXAMPLES OF CONCURRENCY

In programming, and in other activities, concurrency problems can arise when an activity involves several people, processes or machines proceeding in parallel. We will give several examples of concurrency, beginning with one that does not involve computers.

An example: activities in a large project. A large project such as the construction of a building is accomplished by many workers carrying out different tasks. These tasks must be scheduled, and one method of doing this uses *precedence charts,* as shown here.

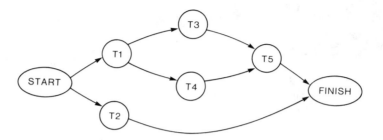

This example chart shows that in the beginning tasks T1 and T2 can both be started. After T1 is done, T3 and T4 can be started, and when both of them are done, T5 can be started. The whole project is finished when T2 and T5 are done. As the next example will show, precedence charts can be used to specify concurrency in computer programs.

There are two main reasons for using parallel tasks in this example. First, there are many workers available and they must be allowed to work at the same time (in parallel). Second, the project can be completed in less elapsed time if tasks are allowed to overlap. In computer systems, analogous reasons (many asynchronous devices and the need to shorten elapsed time) may result in concurrent programming.

An example: independent program parts. Precedence charts can describe possible concurrency in a computation. The expression $(2*A)+((C-D)/3)$ can be evaluated sequentially (one operation at a time) by finding the product, difference, quotient and sum, in that order. But parallelism is possible, because some parts of the expression are independent, as is shown in this precedence chart.

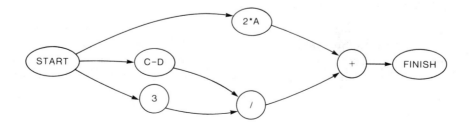

Groups of statements, as well as expressions, may have independent parts that can be executed in parallel. For example, the following loop written in Pascal determines if 'Jones' is in a list by testing name[1] then name[2] and so on.

```
for i := 1 to size do
    if name[i] = 'Jones' then
        found := true
```

This loop could be executed by checking all of the names at the same time, because the tests are independent. For a large computation, parallelism such as this can minimize the elapsed time for completion.

As computing elements such as microprocessors become cheaper, it becomes more and more attractive to split programs into several parallel tasks. In the future we may find that computers are built as huge collections of tiny processing elements, analogous to building an elephant out of a swarm of mosquitoes or bees, and we will need to know how to program such contraptions.

An example: a simulation. Sometimes programs are written to simulate parallel activities. For example, a program might simulate boats entering a harbor; this program could predict the effects of increased boat traffic. A good way to program this simulation is to have an asynchronous program activity (a *process)* corresponding to each simulated activity (each boat). Each process mimics its boat, and the interaction of these processes models the interaction of boats entering the harbor. Programming the simulation is done by writing the constituent concurrent programs.

An example: control of external activities. Special purpose computer systems are used to control chemical processes such as the manufacture of cement. Sensors transmit signals to the computer to report temperature, pressure, rate of flow, etc. The computer in turn transmits signals that set valves, control speeds, sound alarms, etc. The computer system also keeps a log of its actions and prints reports. A computer system such as this keeps track of many interrelated concurrent activities. One good way to program such a system is to have a concurrent software process in the computer for each external activity. A software process tracks its corresponding activity; it is responsible for sending and receiving signals to and from the activity. Programming this computer system is done by writing the concurrent programs that observe and control the activities.

These examples have given various practical uses of concurrency. One of the most important examples of concurrent programming arises in operating systems. The next sections explain why this concurrency arises and how it is handled.

OPERATING SYSTEMS

Modern computer installations have many asynchronous hardware components, such as operator consoles, card readers, printers, disk drives, tape drives and CPUs. The operating system must ensure that these components are used efficiently and that they provide convenient service for the users.

An operating system consists of a collection of software modules. These modules receive requests from users (for example, to execute the users' programs) and must schedule the system's components to satisfy these requests.

The operating system may support *multiprogramming,* that is, it may allow more than one user's program to be in execution at a given time. To support multiprogramming, the operating system must share the system's resources among the executing programs. Some resources, such as tape drives, are exclusively allocated to a program, until the program terminates or no longer needs the resource.

Other resources, such as the CPU, are shared dynamically, in a way that gives the appearance that each program has its own *virtual* resource. For example, the operating system may allocate a "slice" of CPU time to one program, then to another program, and so on. This is called *time slicing* and gives the appearance that each program has a virtual CPU, which is like the physical CPU but somewhat slower. As a second example, the operating system may provide each program with a virtual memory. This is done with the help of special hardware (for "paging" or "segmenting") that allows the operating system to allocate physical memory only to the active parts of programs.

There are two basic reasons why multiprogramming is needed in computer systems. The first is for efficient use of hardware resources and the second is for quick response to users' requests. First we will consider efficiency. The system's hardware components run in parallel at vastly different speeds. For example, the time to process a single character may vary from a tenth of a second for a slow console, to a thousandth of a second for a printer, to a millionth of a second for a CPU. Clearly, the CPU should not be forced to waste time (100,000 of its operations) while a console transmits a character. While a user is typing messages to a running job, another job should be given the CPU. If a job is *I/O bound,* spending most of its time waiting for input/output devices, the spare CPU time can be used by a *compute bound* job, which spends most of its time using the CPU. If the system has a variety of equipment, a job that uses only a few of the devices should not prevent concurrent use of other devices. These examples show how multiprogramming provides more efficient use of computer equipment.

Apart from efficiency, multiprogramming allows the computer system to respond quickly to users' needs. Suppose a user has a short, urgent job but a long-running job is already in the system. With multiprogramming, the short job can run in parallel with the long one and can finish hours before it. In interactive systems, a form of multiprogramming is necessary, with one program for each user. The system is shared among the interactive users and their programs so that each receives good response; this is called *time sharing*. These examples show how multiprogramming allows prompt attention to users' needs.

COMMUNICATION IN OPERATING SYSTEMS

Operating systems must be organized so as to control hardware devices and run users' programs. This section gives a simplified model of how communication occurs among the devices, the operating system and the users' programs; the next two sections describe how operating systems are organized to handle this communication.

An operating system controls an I/O device by sending it a *start I/O command*. If the device is a tape drive, the command may cause it to read a record, putting the record's characters into main memory. When the device has carried out the command, it can send an *interrupt* signal back to the CPU indicating that it is free to carry out another operation. This signal can switch the CPU from a user's job to the operating system; this allows the operating system to send another command to the device before returning the CPU to a user's job.

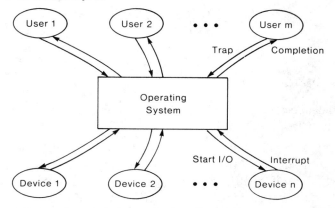

Meanwhile, each user's job occasionally makes requests to the operating system, for example, to read from a terminal or to write a disk track. The job makes a request by a *trap* or a *system call* instruction; this instruction is like a subroutine call and transfers control from the user to the operating system. Having received such a request, the system blocks the job (gives it no more CPU time) until the requested action has been completed. Then the job is unblocked and allowed to continue executing.

There is usually an *interrupting clock;* it sends interrupt signals that transfer control from a user's job to the operating system. This allows the system to implement time slicing by passing the CPU from user to user, and to cancel a user's job that is using excessive CPU time. Without the interrupting clock, an infinite loop in one user's job could prevent other users (and the operating system) from using the CPU.

When a program is actually using the CPU, we say it is *running.* When it is waiting for a request to be serviced, we say it is *blocked.* When a program would be running except that the CPU is allocated to another program, we say it is *ready.* The operating system maintains a queue of the programs that are ready. This transition diagram shows how the states of a program change:

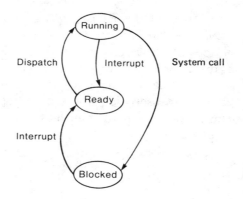

We say the operating system *dispatches* a program when it lets it run, by giving it the CPU.

A trap generally causes a program to be blocked; however in some instances (not shown in the diagram) if the operating system can immediately satisfy the request, the user program is again dispatched and no blocking occurs. Other than by a trap, the only way a running program loses the CPU is by an interrupt. A clock interrupt may signal the end of the program's time slice, or an I/O interrupt may allow another program to run. In a system with only one CPU, at most one program can be in the running state, but with several CPUs there can be as many simultaneously running programs as there are CPUs.

OPERATING SYSTEMS AND MONOLITHIC MONITORS

There is a particularly simple method of handling asynchronism, and of building operating systems, based on the concept of a *monolithic monitor*. Essentially this technique implies that the users' programs and the devices do not communicate directly and that all their interactions are passed through an operating system that cannot be interrupted. We will now explain this idea in more detail.

The interrupt signals are provided by hardware; they allow the operating system to give immediate attention to changes in device status so the devices can be kept usefully active. When the operating system is engaged in some critical activity, for example, updating the queue of ready programs, it must not be interrupted, because the interruption might cause another update of the half-updated queue. The result could be a hopelessly tangled set of pointers and subsequently a system crash. To handle these critical situations, a CPU usually provides a method of disabling or masking *interrupt signals*. When interrupts are *disabled* the hardware holds the signals pending until interrupts are again *enabled*.

The monolithic monitor approach uses disabling/enabling in the following way. Every trap (from the user to the operating system) and every interrupt (from a device or the clock to the operating system) immediately disables interrupts. This means that whenever the operating system begins executing, interrupts have been masked. They remain masked until the operating system dispatches a user program. As a result the operating system is never interrupted, and it gives up control only by handing the CPU to a user program (by dispatching).

The beauty of this approach lies in its simplicity and in the straightforward handling of asynchronism. It is quite easy to build a *very small* operating system on this basis that correctly schedules devices and manages users. But a monolithic monitor is not suitable for most operating systems because of two fundamental problems.

The first is that activity in any part of the operating system disables interrupts from all devices. This means that devices are held up waiting for new commands. For computer systems of any appreciable size, this loss of response to devices can not be tolerated. For certain devices, information is lost if the response is too slow. Inevitably some method must be introduced to allow parts of the operating system to be interrupted, although with great care to avoid scrambling critical data.

The second problem is that tables must be maintained in the monolithic monitor to record the status of each and every device. For example, if a user requests a disk read, the system must know from its tables whether the disk is busy. The size and complexity of these tables can become excessive.

The concept of a monolithic monitor underlies a variety of operating systems. In too many cases the system has outgrown the concept; to improve performance more and more activity is moved outside of the uninterrupted monitor. This gradual evolution of an operating system tends to lead to system errors, as ad hoc tricks are used to try to maintain the consistency of critical shared data. These problems suggest that the responsibility for the various devices should be decentralized and moved out of the monolithic monitor in the initial design.

There is an alternate approach that recognizes these difficulties; it minimizes the system's uninterrupted activities and concentrates these in a module called the kernel.

BASING AN OPERATING SYSTEM ON A KERNEL

The alternative to a monolithic monitor moves the various activities in an operating system out into asynchronous, interruptable software entities called *processes*. Each process is similar to a user's program in a multiprogramming system, in that it shares the CPU's time and is in one of the states: running, ready, or blocked.

One elegant way to structure the operating system is to have a software process (a device manager) corresponding to each device. This manager process has sole responsibility for starting the device and observing when the device becomes free; this means the software process tracks the hardware device. This approach simplifies the operating system's tables because the status of a device is implicitly given by the status of its manager process. For example, a tape drive device is busy exactly when its manager has sent it a command and has not yet been notified of the completion.

Each user's program can be managed and executed by a process. When the the program has an I/O request, it is sent by the user's process to the corresponding manager process. The user's process must then be blocked until the manager notifies it that the I/O has been performed.

Of course there must be mechanisms for *interprocess communication,* so the users' processes can transmit requests and wait for completions. There must also be a mechanism for a device manager to start up its device and to wait for its interrupt, and a mechanism to share CPU time among all the software processes. All these responsibilities are absorbed by a fundamental operating system module called the *kernel.*

The kernel can be implemented like a very small monolithic monitor, which schedules the CPU among the processes, receives I/O interrupts and transmits them to device manager processes, and supports interprocess communication. The kernel uses clock interrupts to help it share CPU time among the processes.

Outside the kernel all interrupts are *invisible* and do not affect the logical progress of the processes. An interrupt can slow down a process by causing its CPU to be temporarily removed, but this is the only effect on the process. The device manager processes observe I/O interrupts, but these appear to the managers as synchronous replies from queries to the kernel, not as asynchronous signals.

The kernel gives each process the appearance of having its own CPU. Each of these *virtual* CPUs behaves like a real CPU except that it has a variable rate of progress. This rate is determined by interrupts, the kernel's CPU scheduling policy, and the use of CPU time by other processes.

The kernel's responsibilities can be implemented by hardware or microprogramming, with the efficiency advantage of making interprocess communication and process/device communication quite fast. Such an implementation has the structural advantage of making asynchronous interrupts invisible to all software modules; this simplifies system design.

Two extremes in organizing operating systems are a huge monolithic monitor versus a minimal kernel that supports only one type of process. Many variations lie between these extremes, and these are greatly influenced by the specific nature of the computer's architecture. Often the processes that manage devices are quite different from those that execute user jobs; on CDC 6000 systems the device managers do not even run on the CPU - they use special "peripheral processors". Use of multiple CPUs, called *multiprocessing* for historical reasons, does not necessarily cause difficulties. In the case of a monolithic monitor, the only additional complication of multiple CPUs is guaranteeing that only one CPU at a time can enter the monitor. Since the kernel in a kernel-based system is implemented as a very small monolithic monitor, the same technique (limiting entry into the kernel to one CPU at a time) applies. Processes (and user programs) outside of the monitor or kernel have their own virtual CPUs, so the presence of multiple CPUs should speed up execution without affecting system correctness.

AN EXAMPLE OPERATING SYSTEM

The T.H.E. operating system nicely illustrates a system based on a kernel. It is a small system that multiprograms several users' Algol programs. It has a hierarchical organization with five levels. The lowest level is the kernel, which implements the concept of processes. In the T.H.E. system, the kernel provides interprocess communication by operations on "semaphores". These operations, called *synchronization primitives,* give each process the ability to block itself and wake up other processes, and the means to guarantee mutually exclusive access to data shared among processes. Semaphore operations will be discussed in detail in the next chapter.

This table lists the functions of the levels in the T.H.E. system:

Level 5	Job managers	Read control language and execute users' programs
Level 4	Device managers	Handle devices and provide buffering
Level 3	Console Manager	Implements virtual consoles for the above processes
Level 2	Page Manager	Implements virtual memories for the above processes
Level 1	Kernel	Implements a virtual CPU for each process

Level 2 of the T.H.E. system is a process that manages a drum and implements automatic paging for all the other processes. After level 2 each process has a large virtual memory which is implemented by automatically moving pages between main memory and the drum.

Level 3 is a process that manages the system console and enables processes on higher levels to communicate with the operator. Level 4 consists of one process per remaining I/O device; there are manager processes for each of the system's readers, printers, plotters and so on.

Finally, level 5 consists of one process per allowed user program. Each of these processes is a job manager. When a job manager completes a user job it searches for a reader that is not occupied. It then reads the control language for another job from the reader and proceeds to execute that user's job. Typically this means compiling the user's program and running it. The job manager is responsible for reserving any I/O devices needed by the job and eventually releasing them.

PROCESSES, PROCESSORS AND PROCEDURES

We have presented operating system structures to demonstrate the necessity of concurrent programming. The concept of a kernel allows us to use concurrent processes even though there is only one CPU (or only a few CPUs). Before going further, we should define precisely some of the basic terminology of concurrent programming.

Process. A process is an asynchronous activity. It can be thought of as the execution of a program by a CPU. However, the CPU may actually be a virtual CPU that is implemented by multiplexing one or more physical CPUs among many processes. A process is, in general, guaranteed to progress through its computation unless it is explicitly blocked. However, its rate of progress may vary considerably.

Processor. A processor is a physical (hardware) mechanism that executes instructions, proceeding from instruction to instruction. A CPU is the prime example of a processor. Some computer systems have special *I/O processors* (or *channels*) whose responsibility is to pass commands to devices.

Procedure. Procedures are sequences of instructions that direct the execution of a processor. Procedures are sometimes called programs. Generally we can separate the data used by a procedure from the procedure itself (the code). If this separation allows the procedure to be executed by more than one process at a time, by setting up a separate data area for each process, we say the procedure is *re-entrant* (or pure). This is analogous to having two cooks simultaneously using the same cookbook (procedure) but using separate pots and ingredients (data).

These three definitions are fundamental to the understanding of both concurrent programming and operating systems. The next chapter begins by giving language features for specifying that several processes are to be executed concurrently.

CHAPTER 1 SUMMARY

This chapter gave several examples of concurrency. Then it presented reasons for using concurrent programming and methods of supporting it. The following important vocabulary was introduced.

Process - an asynchronous activity such as the execution of a program by CPU.

Processor - a hardware mechanism, such as a CPU, that executes instructions, one after another.

Procedure - a sequence of instructions to be executed by a processor; sometimes called code.

Reentrant procedure - a procedure that can be executed by several processes at the same time. It consists of pure code (no writeable data) and each process provides its own data area.

Precedence chart - a diagram that gives the required ordering of several activities.

Multiprogramming - having several programs active at the same time in a computer system. (Each of these activities is a process.)

Time slicing - sharing CPU time among several processes by alternately giving each a short time interval (a slice) of time.

Compute bound - a job that does little input/output but uses a lot of CPU time; an I/O bound job does the opposite.

Start I/O command - a command sent from a CPU (or other processor) to request an operation by a device.

Interrupt - a signal sent from a device to a CPU (or other processor) to indicate that a requested operation is complete. A clock may also send an interrupt to a CPU. An interrupt causes the CPU to switch to execute the operating system.

Trap or system call - a special instruction that a program can execute to switch control to the operating system, for example, when requesting the next input.

Dispatching - giving the CPU to a job so the job can run.

Disabling and enabling interrupts - when a CPU has interrupts disabled, it can not receive interrupts; they remain pending (queued by the hardware) until the CPU is again enabled.

Running, ready and blocked - a process (or job) is running when it is actually using a CPU, ready when it would be using a CPU but none is available, and blocked when it can not use a CPU because the process is waiting, for example, for an I/O completion.

Monolithic monitor - a method of implementing an operating system; all interrupts are disabled whenever any part of the operating system is active. The operating system handles all input/output.

Kernel - a module that implements processes and provides them with a mechanism for interprocess communication. If the kernel is implemented by hardware or microprogramming then the software may not need to deal with interrupts, because device starting/completion is done by the interprocess communication mechanism.

CHAPTER 1 BIBLIOGRAPHY

Hoare gives a brief survey of the function of operating systems. Dijkstra's description of the organization of the T.H.E. system is a classic, well worth reading. Holt's survey of program structures provides a catalog of software structuring mechanisms, including monolithic monitors and kernels.

Dijkstra, E.W. The structure of the T.H.E. multiprogramming system. *Comm. ACM 11,5* (May 1968), 341-346.

Hoare, C.A.R. Operating systems: their purpose, objectives, functions and scope. In *Operating Systems Techniques* (C.A.R. Hoare and R.H.Perrott, editors), Academic Press (1972), 11-19.

Holt, R.C. Structure of computer programs: a survey. *Proceedings of the IEEE 63,* 6 (June 1975), 876-893.

CHAPTER 1 EXERCISES

1. Give a precedence chart that specifies the maximum concurrency for the following program segment. Allow each sub-expression to be computed separately. Be careful not to allow a variable to be assigned a value before its old value is used.

        k := i+7
        j := 7-(5+(3*i))
        i := j+(5*k)

2. Suppose a multiprogramming system is running one compute bound job and one I/O bound job. Which of the jobs should be given priority for using the CPU? Explain why.

3. In certain situations, the throughput of a computer system can be maximized by avoiding multiprogramming and running only one user job at a time. Characterize these situations, taking into consideration the job mix, types of jobs, types and number of peripherals and use of files.

4. As processors (microprocessors in particular) become cheaper, it becomes desirable to decentralize certain operating system responsibilities, moving them away from the CPU and out into channels, devices and terminals. Characterize the types of computational responsibilities that can be moved into each of those locations.

5. Consider a computer system that supports two types of processes: system processes and job processes. Each job process is controlled by a system process. This manager process can suspend (put to sleep) and reactivate (re-awaken) its job process. This results in a new process state that a suspended job process enters, besides the usual three states (running, ready and blocked). Draw the transition diagram showing how job processes can change states.

6. Characterize the sorts of systems in which a monolithic monitor would or would not serve as a good basis for an operating system.

7. Make a list of the visible and invisible interrupts that an executing user job experiences.

8. In the T.H.E. system, if there is a hardware error in reading a memory page, it is impossible for the page manager to notify the operator. Explain why. The solution to this problem seems to be to switch the console manager to level 2 and the page manager to level 3. Explain why such a switch would have a heavy efficiency penalty.

9. Some procedures are "serially re-usable" but not re-entrant. This means that the procedure cannot be used by two processes at once, but can be re-

used after a completed execution without refreshing the procedure. Give examples of program constructs that cause this situation.

10. [E. Lazowska] In Chapter 5 you will be introduced to the problem of the dining philosophers, a well known synchronization problem in which a group of philosophers must cooperatively share forks in an attempt to consume spaghetti. In the following problems we shall consider a more elegant gastronomic enterprise, the Snooty Clam seafood restaurant.

(a) We begin our study of the Snooty Clam restaurant in the kitchen. Earlier in this chapter, we showed how a pair of cooks can be viewed as two processes sharing a common procedure (a recipe). Kitchen facilities (such as the oven) and utensils (such as spoons) can be viewed as resources.

After spending a fortune to create the proper atmosphere in the dining room, the owners of the Snooty Clam found themselves so strapped for cash that they were only able to buy a single saucepan and wire whip. This, of course, means that only one chef at a time can make sauce. The problem is compounded by the arrangement of the kitchen. All pots and pans are hung against one wall, and all stirring implements against another. So in order to acquire the resources required to make a sauce, a chef must visit first one wall, then the other.

Consider the problems that might arise if two chefs simultaneously decide to make sauce, but head for opposite walls in their quest for resources. (Chefs are notoriously temperamental, and refuse to relinquish an acquired resource until they have finished using it.) Devise an allocation policy for the saucepan and wire whip that will prevent the problem you have observed.

(b) The wine cellar in the Snooty Clam lies at the end of a long, narrow tunnel which, unfortunately, is not illuminated. After a recent collision that resulted in three broken bottles, the manager decided that only one of the restaurant's four wine stewards should be allowed in the tunnel at a time, although several can simultaneously use the cellar given that they enter it via the tunnel at different times.

To implement this policy, the manager installed red lights above the entrances at both ends of the tunnel. At each entrance is a switch that turns on or off all these lights. Before entering the tunnel, a wine steward checks to see that the light is off. If so, he first switches it on and enters the tunnel. Upon emerging at the opposite end, he switches the light off.

The manager was very proud of this rather ingenious solution. Unfortunately, though, during its second day of operation a collision occurred in the tunnel. How?

(c) The Snooty Clam restaurant does not take reservations. The dining room contains a single table seating twenty patrons. When space becomes free, parties are seated in the order in which they arrived, except that a party that cannot be seated in the available space is passed over.

What is the effect of this seating policy on large parties? If parties are seated strictly in the order in which they arrive, how will this affect the utilization of the table?

Chapter 2

CONCURRENCY PROBLEMS AND LANGUAGE FEATURES

To solve problems in concurrent programming, we need a good notation for concurrent algorithms. This chapter presents a number of basic concurrency problems and gives programming language features for dealing with them.

SPECIFYING CONCURRENT EXECUTION

To use parallelism in programming we need to be able to specify two or more concurrent activities. The following language feature is sometimes used.

cobegin
    Stmt1
    Stmt2
    Stmt3
    ...
    StmtN
coend

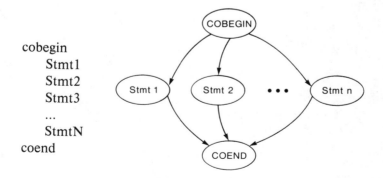

This specifies that the constituent statements can be executed in parallel, as shown in the corresponding precedence chart. Of course, each of these statements may actually be groups of statements. We can think of the cobegin/end construct as creating N concurrent processes, each of which must execute to completion before the creating process is allowed to continue.

There is another notation that can be used to initiate a new activity (a process). It has this form:

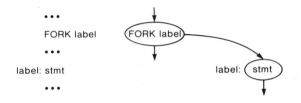

Essentially, the fork statement is a goto statement which simultaneously branches and continues on, as shown in the precedence chart. Fork can be thought of as cobegin without coend. There are additional statements called quit and join that allow the two branches of activity to merge. The newly created activity that started at the specified label executes a quit statement when it has completed its work.

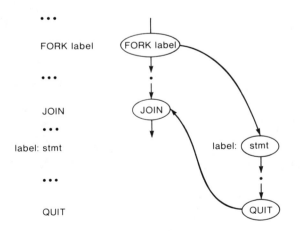

When the original activity, proceeding from the fork, needs to wait until the created activity is complete, it executes a join statement. The process executing join is blocked until the quit has been executed.

We can use fork, join and quit to simulate cobegin/end, just as goto and if statements can simulate a while loop. But cobegin/end is preferred because it is better structured and leads to more understandable algorithms.

DISJOINT AND OVERLAPPING PROCESSES

When concurrent processes use no common data, they are said to be *disjoint* (or independent). Here is a concurrent program to find the maximum of a, b, c and d.

```
1       cobegin
2           m1 := max(a, b)
3           m2 := max(c, d)
4       coend
5       m := max(m1, m2)
```

Since statements 2 and 3 are disjoint, we have no difficulty understanding the program.

When the parallel statements have overlapping data that is changed, things can get confusing. Here statements 3 and 4 are not disjoint.

```
1       j := 10
2       cobegin
3           Print j
4           j := 1000
5       coend
```

At first it appears that statement 3 will print 10, because j is set to 10 in statement 1. But a closer inspection reveals that statement 4 changes j to 1000. If statements 3 and 4 are to be done simultaneously, will j be printed as 10, or as 1000, or maybe as some other value? This presents a problem: just what does parallelism mean when some processes change a variable while others are using it?

The unfortunate answer to this question is that the results depend on relative speeds, and in general we can not predict the speeds of processes. Constructs such as cobegin/end make no guarantee about speeds. Kernels that share time among processes do not control the precise timing of interrupt signals. Hardware processors do not in general execute at precisely defined rates. Even if we could determine the speed of a process for a given execution, each successive execution might be different.

When the outcome of a computation depends on the speeds of processes, we say there is a *race condition*, and that parts of the computation are *time critical*. Operating systems use concurrency to maximize throughput and convenience; but they must be carefully designed to prevent race conditions that could destroy the system or the results of users' programs.

We will give two more examples of processes with overlapping data to illustrate the danger of race conditions. Suppose one process, called the observer, is responsible for observing and counting certain events; for example, it may observe the number of jobs submitted to a computer center. It executes this program:

    Observer: loop
                        Observe an event
                        count := count + 1
                    end loop

Another process, called the reporter, occasionally prints reports about the observed events. The reporter executes this program:

    Reporter: loop
                        Print count
                        count := 0
                    end loop

As soon as the reporter prints the count of events, it sets the count to zero, because the events have been reported.

The observer and reporter are overlapping in that they both use the variable called count. This overlap causes a problem. Suppose that the observer has increased count to 6 and the reporter prints 6. Suppose that before the reporter sets count to zero, the observer increases count to 7. Now suppose the reporter continues and changes count from 7 to 0; the unfortunate result is that an event goes unreported. In general, the reporter may fail to report any number of events because increments to count may occur between printing count and setting it to zero.

There is another problem in this example that is less obvious, and has to do with the statement that increments count. An implementation of the statement count := count+1 may involve more than one machine instruction, such as:

        LOAD   COUNT   (Put COUNT in accumulator)
        ADD   1   (Add 1 to accumulator)
        STORE COUNT   (Store accumulator into COUNT)

When the observer process is executing this sequence, it may be overtaken by the reporter. Suppose count is 15, and this value is loaded into the accumulator. Then the reporter may print 15 and set count to zero. Next the observer adds 1 to its accumulator and stores the result 16 into count. The unfortunate result is that 15 events were reported, but count is left indicating that 16 events are yet to be reported.

From this example we conclude that when processes update shared data, the results can be unpredictable and not at all what is desired. In the next section we will show how to deal with this problem, but first we will give another typical example of race conditions.

In operating systems there are often queues, for example, queues of processes ready to use the CPU. Consider a singly linked queue:

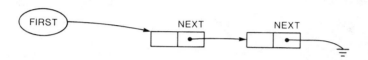

As shown this queue has two elements. To insert a new element e into the queue, the following is executed:

    node(e).next := first
    first := e

This changes the queue to the following:

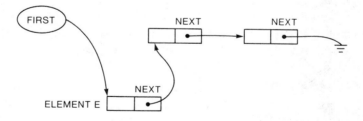

Suppose that while e is being inserted, another process is trying to insert element f by executing

    node(f).next := first
    first := f

Suppose the two processes simultaneously set node(e).next and node(f).next to first. Now suppose the variable first is set to f by one process and then immediately re-set to e by the other. The resulting messed-up queue has this form:

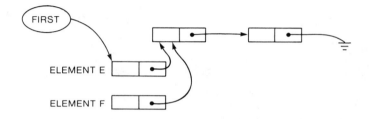

Unfortunately, element f has been lost; it cannot be found by following links from first. This would be disastrous if, for example, element f represented a job to be executed.

The race conditions given in these examples have the same disastrous results whether the processes execute physically or logically in parallel. By *physically parallel* we mean that each process has a processor (CPU) and these are simultaneously running. When processes are implemented by time slicing of a single CPU and only one process at a time can be running, the processes execute *logically* in parallel. In the case of time slicing, race conditions occur as a result of the unpredictable transfer of the CPU from one process to another.

The solution to these race conditions is to make sure that only one process at a time gains access to the shared data that is updated. Each of these accesses occurs in a part of the program called a "critical section"; we will now discuss methods of guaranteeing "mutually exclusive" access to these critical sections.

CRITICAL SECTIONS

As we have seen, when a process is updating variables, it is generally unreasonable to allow any other process to access the same variables. The required control of access can be accomplished by the mutexbegin/end construct, as illustrated here:

Process P1:
  loop
    Compute
    mutexbegin
      Access shared variables
    mutexend
    Compute
  end loop

Process P2:
  loop
    Compute
    mutexbegin
      Access shared variables
    mutexend
    Compute
  end loop

The mutexbegin/end construct, with its brackets mutexbegin and mutexend, guarantees that a process will have *mutually exclusive* access to the sections of programs within the brackets. At a particular time, P1 or P2, but not both, are allowed to execute the part of their program, called a *critical section*, within the brackets. A process is allowed to use several mutexbegin/end constructs, but these cannot be nested.

A generalization of mutexbegin/end would have a parameter specifying a particular set of shared variables. For example, we could have:

  mutexbegin(v)
    Access to shared variables in collection v
  mutexend(v)

But we will keep this discussion simple by ignoring the parameterized construct.

One of the fundamental problems in concurrent programming is how to implement mutexbegin/end. Its two brackets must accomplish the following:

Mutexbegin. Must determine if there is any other process in a critical section: has another passed a mutexbegin but not the corresponding mutexend? If so, the entering process must wait. When no other process is in a critical section, the process proceeds beyond mutexbegin, setting an indicator so that other processes reaching a mutexbegin will wait.

Mutexend. Must allow a waiting process, if there is one, to enter its critical section.

It *seems* easy enough to implement mutexbegin and mutexend. A flag called occupied can be initialized to false to indicate that no process is in a critical section. Then mutexbegin can be written as:

```
1        loop
2            exit when not occupied
3        end loop
4        occupied := true
```

The flag called occupied is repeatedly tested until it is found to be false, then it is set true and the process enters its critical section. Mutexend can be written as:

```
        occupied := false
```

These implementations seem to accomplish our goal of allowing only one process at a time into a critical section. But appearances are deceiving in concurrent programming.

Our implementation is just plain wrong. Suppose process P1 tests occupied in statement 2 and finds that it is false. At the same time, or very shortly before or after, P2 may make the same test. Both processes, executing in parallel, will conclude that occupied is false, will proceed to execute statement 4, and will enter their critical sections at the same time. In the next section we will be more careful when we try to develop a solution for the mutual exclusion problem.

MUTUAL EXCLUSION BY BUSY WAITING

In computer systems with more than one (hardware) processor, the processors may occasionally need to have mutually exclusive access to certain data. When one processor is using the data, another processor wanting to use the data must wait; this waiting can be accomplished by repeated execution of a test to see when the critical section can be entered. This repeated testing is called a *busy wait* and is clearly a waste of processor time. This can be tolerated when the critical sections are used only a small fraction of the time, say five per cent, or when processor time is not considered particularly valuable.

Sometimes busy waiting is necessary; one of the prime examples is in a multiple CPU system. In such a system there is usually a set of queues, including the ready queue, that maintains the status of processes. The kernel manages these queues, and to keep the queues from becoming tangled, only one CPU at a time should enter the kernel. The kernel should be designed to be very fast so that a negligible amount of processor time is lost via busy waiting. In the rest of this section we show how busy waiting and a property of memory called "interlock" can be used to guarantee mutual exclusion.

There is special circuitry that controls the accessing of the memory by processors. This circuitry makes the memory act like a device that receives commands from the processors to fetch and store words or bytes; this memory device carries out only one command at a time. (Systems with multiple banks of memory may support *memory interleaving*, in which each bank carries out only one command at a time.) This "one command at a time" access is called *memory interlock.*

If one processor is the CPU and the other is a channel (an input/output processor) then the channel is usually given priority by the memory circuitry; we say the channel *steals cycles* from the CPU because the CPU must wait (and lose memory cycles) while the channel transfers data to or from the memory.

In the absence of memory interlock we cannot be sure that concurrent execution of the assignment statements $j := 10$ and $j := 1000$ will leave j as either 10 or 1000; j might end up as a random bit pattern. With memory interlock, j ends up as either 10 or 1000, but we may not be able to predict which.

Given memory interlock we can implement mutexbegin/end in the following manner. We initialize a shared variable called turn to either 1 or 2. We use separate local copies of myTurn and hisTurn for P1 and P2; for P1 they are initialized to 1 and 2, and for P2 to 2 and 1, respectively. Mutexbegin is implemented as:

```
loop
    exit when turn = myTurn
end loop
```

And mutexend is:

```
turn := hisTurn
```

This "solution" has an awful shortcoming; it requires that P1 enter its critical section, then P2, then P1 and so on. If P2 is ready to use its critical section first, too bad! It must wait until P1 catches up, and this strict alternation continues.

We will now give an implementation of mutexbegin/end that avoids strict alternation, but still uses busy waiting. We will also use a shared variable called turn that is initialized to either 1 or 2 (to myTurn or hisTurn), and shared variables called need(1) and need(2) that are initialized to false. Each process has local variables me and other; these are initialized to 1 and 2 for the first process and to 2 and 1 for the second. This implementation is called *Dekker's algorithm*, after Dekker, the Dutch mathematician who devised the original version of it.

```
mutexbegin:
    need(me) := true
    turn := hisTurn
    loop
        exit when not need(other) or turn = myTurn
    end loop


mutexend:
    need(me) := false
```

When critical sections are not being used heavily, mutexbegin usually finds that need(other) is false, signifying that the other process has not requested entry. Entry into the critical section is therefore immediate, with no repetitions of the loop. The variable turn is used to decide which process is to enter only when both processes have requested entry.

The logic of Dekker's algorithm is subtle and can be appreciated by attempting to program another solution. The algorithm can be generalized to handle any number of processes, but becomes even harder to understand.

The main point about Dekker's algorithm is that it demonstrates that with only memory interlock and busy waiting, mutual exclusion can be guaranteed. The algorithm satisfies the following requirements for mutual exclusion.

(1) Only one process at a time is allowed in a critical section.

(2) A process will be allowed to enter its critical section if no other process is using a critical section.

(3) No set of timings can keep a process waiting indefinitely as it tries to enter its critical section.

We now demonstrate that Dekker's algorithm meets these requirements. Point (1) would be violated if both processes somehow got into their critical sections, which would set both need(me) and need(other) to be true. Both could not enter by being in their testing loops at the same time, because "turn" can favor only one. So the second process must have set turn to be hisTurn before entering the testing loop, in which case the test fails and mutual exclusion is maintained! So point (1) is satisfied.

Point (2) is clearly satisfied because the entering process finds need(other)=false. Point (3) can be violated only if a process is blocked forever in its testing loop. In this event, the other process is necessarily doing one of three things: (a) not trying to enter, (b) cycling through the testing loop or (c) repeatedly entering and leaving critical sections. Case (a) cannot cause blocking because need(other) is false. Case (b) is impossible because "turn" necessarily selects one process to enter. Case (c) cannot continue because the other process will set turn to favor the first process. So point (3) is satisfied.

There is a simpler implementation of mutexbegin/end, given that the processor has an instruction that both tests and sets (modifies) a word. For example, if the processor has a condition code called oldFlag, the operation could be:

```
1      TestAndSet(flag, oldFlag):
2          oldFlag := flag  {Test the value of flag}
3          flag := true     {Set the value of flag}
```

Lines 2 and 3 must be carried out by a single, uninterruptable machine instruction. If there are multiple processors, memory interlock must prevent another processor from accessing the flag between the test (line 2) and the set (line 3). Note that each processor has its own condition code (oldFlag).

We use a shared variable called occupied that is initialized to false and we implement mutexbegin/end as follows.

```
mutexbegin:
    loop
        {See if occupied is true; set it false}
        TestAndSet(occupied, wasOccupied)
        exit when not wasOccupied
    end loop
```

```
mutexend:
    occupied := false
```

Not only is this solution simpler than Dekker's algorithm, it also handles any number of processes. Even when there is no instruction called "test and set", there may be an instruction with the desired properties. For example, we may be able to use a DECREMENT instruction which subtracts one from a word and sets the condition code if the result is negative.

In this section we used busy waiting to solve the mutual exclusion problem. We developed an unsatisfactory solution that implied strict alternation among processes entering a critical section, and two good solutions: Dekker's algorithm and the test-and-set method. Generally, busy waiting is an unacceptable waste of CPU time. Even when each process has a CPU dedicated to it, and waiting necessarily wastes CPU time, busy waiting may be impractical. The problem is that busy waiting can saturate the electronic path to memory; this *memory contention* slows down all the CPUs. Busy waiting can be avoided by using special operations called synchronization primitives.

SYNCHRONIZATION PRIMITIVES: SEMAPHORES

Constructs such as cobegin/end and mutexbegin/end can be thought of as *primitive operations* (or *primitives*) because they have a simple meaning, and because we can use them without knowing their implementation. One of the best known sets of primitive operations for process synchronization is based on special variables called *semaphores*.

In a programming language, we might declare a semaphore called s this way:

var s: semaphore initial(1)

The only valid operations on semaphores are P (sometimes called Wait) and V (sometimes called Signal). These operations were developed by Dijkstra; P and V are abbreviations of the Dutch words for waiting and signaling. The two semaphore operations allow a process to block itself to wait for a certain event and then to be awakened by another process when the event occurs. P and V have the following meaning.

P(s): Wait until s $>$ 0 and then subtract 1 from s.
V(s): Add 1 to s.

Both P and V must be done indivisibly. The P operation potentially blocks the executing process and V potentially wakes up a blocked process. The process executing the V operation is not blocked and continues execution.

A semaphore can be thought of as a bowl to hold marbles. The numeric value of the semaphore corresponds to the number of marbles in the bowl. The initial attribute in the semaphore's declaration gives the original number of marbles in the bowl. Each executed V operation puts a marble into the bowl. Each executed P operation attempts to remove a marble. If none is available, P causes the process to wait. The process waits until a marble is available, removes the marble, and continues execution.

Typically, semaphores are implemented by a software kernel. Some systems, notably the VENUS operating system, have a kernel that is implemented in micro-code. When a process becomes blocked by the P operation, the kernel allocates the CPU to another process that is ready to run. This allocation avoids busy waiting and so avoids wasting CPU time.

Although busy waiting is not generally practical, we will show how it can be used with mutexbegin/end to implement P and V. In the following, local (separate) variables called blocked are used by each process.

```
V(s):    mutexbegin
             s := s+1
         mutexend


P(s):    begin
             var blocked: Boolean := true
             loop {Busy wait}
                 mutexbegin
                     if s > 0 then
                         s := s - 1
                         blocked:=false
                     end if
                 mutexend
                 exit when not blocked
             end loop
         end
```

Note that the test to see if s is greater than zero must be in the same critical section with the statement that decrements s; otherwise two P operations might erroneously decrement s when its value is found to be 1. This would correspond to two processes erroneously grabbing the same marble from the semaphore bowl.

Given that semaphores are available - and that we can assume they have been implemented by the kernel - we can use them for synchronizing processes. If two processes want to update the same variable without interference, they can use a semaphore, which we will call mutex.

```
var mutex: semaphore initial(1)
```

Before updating the critical variable, a process executes:

```
P(mutex)  {Implements mutexbegin}
```

After the update, the process executes:

```
V(mutex)  {Implements mutexend}
```

The semaphore called mutex originally "contains a single marble" because of initial(1) in the declaration. The P operation removes the marble and V puts it back. When a process is updating the critical variable, the marble is gone, so any further P operations are blocked until the update is complete.

Notice that this implementation shows that semaphores can be used to solve the mutual exclusion problem. If there are several independent critical variables (or critical data structures or resources) then a semaphore can be declared for each one to provide separate mutually exclusive access.

Semaphores can also be used to provide processes with the *block/wakeup facility*, which allows each process to wait until certain events occur. For example, suppose process R must wait until process Q has completed a certain action.

Q:Compute	R:Compute
Wakeup process R	Block until awakened by Q
Compute	Compute
...	...

The wakeup and block operations can be supported by having a *private semaphore* for each process, provided by a vector privateSem of semaphores initialized to zero. Each process has a distinct process number, and this is held in the process's local variable called me. The block/wakeup operations are implemented as:

> Wakeup(processNo):
> > V(privateSem(processNo))

> Block:
> > P(privateSem(me))

Each private semaphore corresponds to a bowl that initially has no marbles in it. The wakeup operation deposits a marble in the bowl, and the block operation attempts to remove a marble. Notice that this implementation works in our example when Q does the wakeup before R blocks and as well when R blocks before Q does the wakeup.

We gave implementations of block/wakeup and mutexbegin/end that use semaphores requiring only the values 0 and 1, corresponding to zero or one marbles. Such semaphores are called *binary semaphores*. They are somewhat simpler than *general* or *counting semaphores*, whose values can be any non-negative integer.

Block/wakeup and mutexbegin/end are sufficient by themselves to synchronize processes in an operating system. Since these operations can be implemented by binary semaphores, binary semaphores are also sufficient. Although binary semaphores are sufficient, they are not particularly convenient or well structured; for this reason this book will concentrate on a more sophisticated synchronization method based on monitors. But before introducing monitors, we will complete our discussion of synchronization primitives such as semaphores.

To illustrate the use of binary semaphores in solving synchronization problems, consider a set of processes that share a resource having several identical units. The units might be, for example, tape drives or data buffers. When a process needs a unit (or another unit) of the resource, it executes Request, and when done with a unit, it executes Release. The Request operation blocks the process when all the units are already allocated. To implement these operations we will use a variable called avail that is initialized to the total number of units, is decremented by Request, and is incremented by Release. Here is the implementation:

Request:
    mutexbegin {Implemented by P(mutex)}
        blocked := (avail=0)
        if blocked then
            Put integer me on queue to wait for a unit
        else
            avail := avail - 1
        end if
    mutexend {Implement by V(mutex)}
    if blocked then
        Block {Implemented by P(privateSem(me))}
    end if

Release:
    mutexbegin {Implemented by P(mutex)}
        avail := avail + 1
        Determine u such that process u is waiting for a unit
        if waiting process u exists then
            Remove u from queue waiting for a unit
            avail := avail - 1
            Wakeup(u) {Implemented by V(privateSem(u))}
        end if
    mutexend {Implemented by V(mutex)}

This implementation can clearly be done with binary semaphores because the only synchronization operations used are for mutexbegin/end and block/wakeup. Interestingly enough, this implementation demonstrates that counting semaphores can be provided using binary semaphores; this follows from the observation that Request and Release behave just like P and V and the value of avail corresponds to the value of a counting semaphore. If we had counting semaphores available, the Request and Release operations could be implemented simply as P(r) and V(r) where r is a semaphore initialized to the total number of units.

OTHER SYNCHRONIZATION PRIMITIVES

There are other primitive operations, besides P and V, that have roughly the same capabilities. For example, the *lock* and *unlock* primitives are used in some systems to guarantee mutually exclusive access to a particular object or data structure. A programming language could provide "gates"; for example, a declaration might appear as:

　　　var g: gate

We can consider that a shared object is surrounded by a fence with a gate in it. Before using the object, a program should lock its corresponding gate, and afterwards should unlock it:

　　　Lock(g)　{Similar to mutexbegin}
　　　Unlock(g)　{Similar to mutexend}

These two primitive operations are like the parameterized version of mutexbegin/end. A variation of lock/unlock used in IBM 360/370 operating systems is called enqueue/dequeue; these allow specification of either shared or exclusive access. Exclusive access is just what we have already discussed. Shared access means many processes at a time can access the object, typically for read-only usage. Enqueue/dequeue prevent overlap of shared and exclusive access.

Lock/unlock and enqueue/dequeue provide a generalized version of mutexbegin/end. Other primitives provide a facility similar to block/wakeup. These are often based on synchronization variables called "events", which are declared in PL/I as:

　　　DECLARE E EVENT;

The declaration implicitly initializes E to false. In PL/I the following statements operate on events.

WAIT(E);	Blocks process until E is true
COMPLETION(E) = TRUE;	Sets event E to true
COMPLETION(E) = FALSE;	Sets event E to false

Events in PL/I are different from semaphores in that the wait operation does not change the value of the event; by comparison, the P semaphore operation decrements a semaphore. The lack of change of the event's value by wait is inconvenient in the common case of repeated use of an event; this situation requires explicit resetting of the event to false after the process wakes up, and this must be done with great care to avoid losing the next wakeup signal.

Events can be used in PL/I to provide a facility similar to fork/quit/join. A new process can be started up by executing:

CALL P EVENT(E);

This is like a fork operation and it starts up a new "child" process executing procedure P. The calling process can be thought of as the parent. When the child process has completed its job, it can terminate by executing an EXIT statement. This sets E to true, so the parent can wait for the completion by executing:

WAIT(E);

The child's EXIT statement corresponds to QUIT and the parent's WAIT(E) corresponds to JOIN. As we will see in a later chapter, the Unix operating system supports this kind of WAIT and EXIT.

We will now leave primitive synchronization operations, as typified by P and V, to consider more sophisticated interprocess communication methods such as message passing. In some operating systems a method such as message passing is directly supported by the kernel; in others, the kernel supports only primitive synchronization operations and these may be used to implement the more sophisticated operations.

MESSAGE PASSING

One of the principal uses of synchronization primitives is to allow processes to exchange information. This fact suggests that these primitives should be generalized to become communication operations that provide both synchronization and data transmission.

The Send and Receive operations are such mechanisms. A process executes Send to pass a *message* (some information) to another process; the other process accepts the information by executing Receive. For example, suppose a user process wants to have the disk manager read a certain track from the disk. The user might execute:

Send command to disk manager

The command specifies the desired disk input/output operation. The disk manager executes this program:

loop
      Receive user command {Wait for next command}
      Start up the disk to carry out the command
      Wait until the disk is finished with the command
      Send user the status of command
   end loop

A more complicated manager might accept and queue several requests while the disk is busy. When the user program wants to wait until the command is complete, it executes:

Receive status of command

This blocks the user until the command is complete and the manager sends the status.

There are many different types of Send and Receive operations. These differences arise from the various methods of storing messages and routing them from sender to receiver. In most systems the sender of a message continues executing after executing Send. This implies that the message must be stored until received in some special place, and this place is called a *mailbox*. In a programming language, mailboxes called c and s might be set up by:

var c: mailbox
var s: mailbox

The mailbox called c could hold commands and s could hold status. Assuming each process has local variables called command and status, the sends and receives of our example could be:

Send command to c   {User puts command in mailbox c}
Receive command from c   {Manager gets command}
Send status to s   {Manager puts reply in mailbox s}
Receive status from s   {User gets reply}

The programming language TOPPS has constructs that are equivalent to this type of send/receive.

One of the problems with mailboxes is deciding upon an appropriate size for messages. The simplest solution is to force all messages throughout the system to have the same length. The RC4000 operating system takes this approach and uses a system-wide length of 24 characters. Messages that are shorter than 24 characters must be padded out to the required length. Some "messages", such as input/output transfers, will be too long to fit in a mailbox. This difficulty is called the *large message problem*; it is solved by using messages to transmit control information (such as input/output commands) and using a separate mechanism to transfer the large amount of data.

The mailbox scheme designed for the SUE/360 operating system is extremely general. It allows *multiple-slot* mailboxes, which can hold several messages (one per slot). Mailboxes are created dynamically and the size of a mailbox's slots are specified at creation. The slots are arranged so that messages are received in the order they are sent (first-in-first-out). Each mailbox has an input port for accepting messages, and an output port for transmitting messages. Each process has a number of input ports and a number of output ports, attached to corresponding mailbox ports. This rather elaborate scheme was never implemented. Instead, a simpler mechanism was adopted; the adopted scheme is similar to monitors, which will be presented in detail later.

The Unix operating system provides an elegant form of mailboxes. These are called "pipes" because they provide a channel that streams data from one process to another. Suppose a person types the command "ls" on a Unix console. The "ls" stands for "list" and causes the names of the files in the user's directory to be printed by the console. The flow of information is from the directory via the "ls" program to the console:

The user can specify that the list of names is to appear on the offline printer instead of the console by typing:

ls |opr

"Opr" is a program that takes a stream of information and prints it. The symbol "|" specifies that a pipe is to channel the output of "ls" so it becomes the input to "opr". The flow of information now becomes:

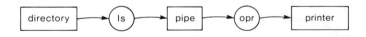

In Unix all transmission to/from devices, files, and pipes is done by a standard set of operations called read and write. The number of bytes to be transmitted is specified in each operation. These read/write operations can be considered to be special forms of receive and send. They provide a powerful software tool by allowing software modules to be connected easily in various configurations. Each module is written without consideration of whether its input or output is to or from a device, file, or process. Later chapters discuss Unix and its pipes in more detail.

Whenever one process produces a sequence of outputs that is used by another process as input, we say there is a *producer/consumer relationship.* In the Unix example, the "ls" program produces lines that are consumed by the "opr" program. In general, mailboxes and message passing are used to facilitate such relationships.

There are some simplifications that can make message passing more efficient than is possible for an elaborate scheme such as that proposed for the SUE/360 operating system. For example, the system may support only *single-slot* mailboxes; this simplifies the administration of mailboxes, but decreases the ability of the mailboxes to provide buffering against differing speeds of the sender and receiver. Ultimately, all the slots of a mailbox can be eliminated, leaving only a "zero-slot" connection between the processes. In this arrangement if the Send occurs first, the sender is blocked until the Receive occurs; then the transmission of the message takes place and both processes are allowed to proceed. Conversely, if the Receive occurs first, the receiver is blocked until the Send occurs. Another simplification is to shorten all messages to zero characters (to null messages) and to use Send and Receive only for synchronization - not for communication. When this is done, receive and send revert to being synchronization primitives similar to P and V.

Message passing has been used as a structuring tool for handling asynchronism in various operating systems, notably the RC4000 system and the Thoth system. However, unless great care is taken, the Send and Receive operations are too slow, commonly requiring on the order of 1000 machine instructions each. Besides, mailboxes tend to impose a network structure on an operating system, with each process being a node connected by mailboxes to other nodes. Such a network does not encourage the hierarchical organization of an operating system.

THE BLOCKING SEND

Message passing can be simplified if the sending process is blocked until it receives an answer to its message. The idea is to combine Send and Receive into a new operation, which we will call BlockingSend. In terms of our example of a user process that sends a request to the disk manager, the user executes

    BlockingSend command, status to manager

This sends the command as a message to the manager; the user process is blocked until the manager sends back a message giving the status.

The manager executes the same program as before, but now we rename its Receive as Accept, and its Send as Reply:

```
loop
    Accept command {Wait for next command}
    Start up disk to carry out command
    Wait until the disk is finished with the command
    Reply with status
end loop
```

The "blocking send scheme" eliminates Send and Receive and replaces them by the three new operations: BlockingSend, Accept and Reply. Accept can only receive a message sent by BlockingSend, and Reply can only answer a message received by Accept.

This scheme offers various advantages. A process that initiates a conversation (by BlockingSend) cannot start a new conversation until it receives a reply. Therefore we can allocate the storage for each message and reply as a part of the initiating process's data. This answers several problematic questions about how to implement mailboxes, such as how many slots to provide and how large to make slots. Essentially, the message and reply are parameters to the procedure BlockingSend. Another advantage of the scheme is that it is easy for the manager to know where to send its reply. The reply obviously goes back to the sender, and the sender's address can be implicitly included as a part of the message.

THE RENDEZVOUS

The Ada language takes the blocking send scheme a step farther with its language construct called the "rendezvous". In Ada, a process that wishes to behave as a manager has "entry points" corrresponding to each kind of request that it is willing to carry out. Our disk manager accepts only one kind of request; its only entry point could be declared by:

```
entry DiskRequest(command: in CommandType,
        status: out StatusType)
```

The keyword "in" means the command comes into the manager and "out" means the status is passed back out.

To request a disk operation, the user executes:

```
DiskRequest(command,status)
```

This appears to be a procedure call, and behaves like one in that the caller is blocked until the requested action is complete. The manager now has the form:

```
loop
    accept DiskRequest(command: in CommandType,
                       status: out StatusType) do
        Start up the disk to carry out command
        Assign result to "status"
    end
end loop
```

This loop contains an "accept" block of the form:

```
accept ...entry name and parameters... do
    ...carry out request...
end
```

The first line of the block, from "accept" to "do", has the same meaning as our previous Accept operation. But now we do not have a Reply operation because there is an implicit completion of the request when we reach the "end" of the accept block. We say the user has a "rendezvous" with the manager; the user process remains blocked from the time it executes DiskRequest until the manager reaches the "end" of the accept block. The manager is blocked at "accept" until the user executes DiskRequest. This is similar to the BlockingSend scheme except that now the effect of Reply is syntactically attached to the keyword "end".

One obvious difficulty with the rendezvous is that a manager does not in general wish to reply to requests in the same order in which they are received. For example the manager may wish to receive several requests and order them to optimize disk performance; the requests will not be completed in FIFO order. Unfortunately, Ada's accept blocks imply FIFO ordering.

To write a manager in Ada that re-orders its requests requires an involved program. The manager has an entry that accepts requests, but immediately replies, telling the user to submit a new request to another entry point. The user is queued at the other entry point while the manager carries out the request. The manager must keep track of these pending requests, and must eventually rendevous with waiting users at these other entry points.

A difficulty with blocking sends and with rendezvouses is that the manager may wish to conditionally accept a request, or possibly to re-queue a request that cannot be immediately handled. Some versions of blocking send schemes allow re-queueing. Ada supports conditional rendezvouses by means of a "select" statement that resembles an elaborate "case" statement:

```
select
    when ...Boolean expression... = >
        accept ...entry name and parameters... do
            ...carry out request...
        end
        ...
    when ...Boolean expression... = >
        accept ...entry name and parameters... do
            ...carry out request...
        end
        ...
end select
```

This "select" statement contains several "accept blocks" each of which is preceded by a "guard" contained in the form:

```
when ...Boolean expression... = >
```

The manager executes a select statement to conditionally accept requests. The Boolean expressions (guards) are evaluated in random order. Each one that is true allows its corresponding "accept" to be executed, given that a user process is attempting to rendezvous at that entry point. Only one of the accept blocks will be executed.

The idea of a "guard" to implement conditional acceptance of messages was introduced in "communicating sequential processes", which will now be described.

COMMUNICATING SEQUENTIAL PROCESSES

Hoare introduced a notation called CSP (communicating sequential processes). The notation is intended to facilitate the combination of concurrent algorithms into larger programs. CSP uses a blocking send, whose form is:

```
processName ! message
```

The executing process sends a message to the named process, and is blocked until the message is received.

The receiving process executes a "receive" operation of the form:

```
processName ? message
```

The receiving process is blocked by this statement until the sender executes the "!" statement.

The scheme is similar to an Ada rendezvous between sending and receiving processes. However, it is different from a rendezvouz in that the sender is unblocked as soon as the message is transmitted to the receiving process. It is also different in that a message is sent to a process rather than to an entry point of a process.

Message passing in CSP uses a "zero slot" mailbox, meaning that no storage is necessary to store the message between the time it is sent and received. This is because the message is copied directly from the sender's storage to the receiver's storage.

The CSP notation uses special characters, such as [, instead of key-words such as "if". To give an idea of how CSP is used, we will give an example process. This process reads characters from a process called "west" and writes characters to a process called "east". The stream from "west" is simply copied to "east", except that each pair of adjacent asterisks is translated to a single "up arrow" character. The CSP notation is given on the left and more traditional notation is given on the right.

```
     *[                                    loop
          west?c ->                             Receive c from west
          [ c≠asterisk ->                       if c not= asterisk then
               east!c                                Send c to east
          |c=asterisk ->                        else
               west?c                                Receive c from west
               [ c≠ asterisk ->                      if c not= asterisk then
                    east!asterisk;                        Send asterisk to east
                    east!c                                Send c to east
               |c=asterisk ->                       else
                    east!upArrow                         Send upArrow to east
               ]                                     end if
          ]                                     end if
     ]                                    end loop
```

In CSP, a "guard" appears between [and ->; this guard determines the set of alternatives which can be executed. For example, on the second line, "west?c" is a guard which blocks execution until a character is received from the "west" process.

One difficulty with the CSP scheme is that messages must be sent to particular processes. Consider a manager process that serves several custo-mer processes. Any message from a customer process to the manager pro-cess must include the customer process's name. Otherwise the manager would not know what customer to reply to.

We now consider a different scheme, called monitors, that overcomes many of the shortcomings of message passing.

MONITORS

The concept of a monolithic monitor, discussed in Chapter 1, can be generalized into a programming language feature for handling synchronization. The feature is called a *monitor* and provides convenient facilities for guaranteeing mutual exclusion and for blocking and waking up processes. We will now give an introduction to monitors; later, in Chapter 4, we will cover monitors in detail as a feature of the language Concurrent Euclid.

Previously the lock/unlock synchronization primitives were described in terms of a fence around critical data; a gate in the fence controls access to the data. The code executed between lock and unlock corresponds to the critical section for that particular data. A monitor can also be thought of in terms of a fence around critical data. One difference from lock/unlock is that all sequences of statements that manipulate the data are collected and moved inside the fence. The fence has several gates, one corresponding to each sequence of statements. Each of these sequences becomes a special purpose procedure called an "entry". This means that all the critical sections for a particular set of shared data are collected into one place.

Whenever one of these entries is invoked, exclusive access to the shared data is automatically provided, so only one process at a time is allowed inside the enclosure. The enforcement of mutual exclusion is implicit: the programmer needs only to invoke the entry. Given that monitors are a construct in a high level language, it is up to the language translator to generate code to implement mutual exclusion.

Monitors provide a block/wakeup facility in the following way. If a process enters a monitor and finds that a required condition (such as the availability of a free resource) is not true, it executes a wait statement. This removes the process from the monitor, blocks its progress, and places it on a queue waiting for the condition to become true. When another process enters the monitor and finds the condition to be true, it executes a signal statement that removes a waiting process (if there is one) from the condition's queue and wakes it up.

The following illustration shows a monitor with three gates, for entries E, F and G, and two conditions, C and D. There is one process, P5, inside the enclosure; P5 has entered gate E to access the critical data. Processes P1 and P2 are blocked at gate E and P4 at gate G; these three processes must wait until no process is inside. Processes P3 and P6 have executed wait statements for condition C. Condition D currently has no processes waiting for it.

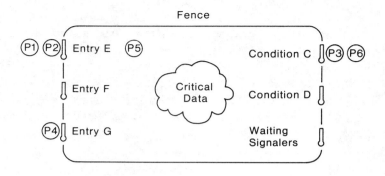

If process P5 now leaves the monitor, one of the processes P1, P2 or P4 will be allowed to enter. If, instead, P5 executes a signal statement for condition C, one process, either P3 or P6, will be allowed to enter. Assuming P3 enters, the situation changes to this:

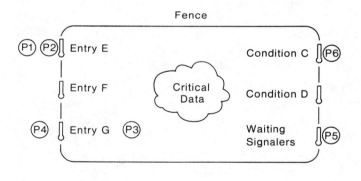

We are assuming that P3 was in entry G when it executed a wait for condition C. The signal by P5 causes P3 to continue executing in entry G. Since only one process at a time is allowed inside the enclosure, P5 is forced to step outside while other processes are inside. If P3 next signals condition C, then P6 enters the enclosure and P3 steps out to join P5. If a process signals a condition that has no processes waiting for it, such as D, then nothing happens (good or bad). However, the signaling process may temporarily step out of the enclosure even when there are no waiting processes.

In Hoare's formal definition of monitors, he assumes that the critical data for each monitor has associated with it a consistency criterion. This criterion is called an invariant, or I for short, and means that the data should be accurate and up-to-date. For example, the critical data used in allocating a particular resource should accurately represent the current status of the resource. A process enters the monitor to test or update the critical data, but must always leave the data so it corresponds to the resource's current status. When the monitor is created, the initial status of the resource is recorded in the critical data, so the invariant I is initially true. Before a process leaves an entry or signals a condition, it must make sure that the critical data is consistent and up-to-date, so I is left true. As a result, any process that enters the monitor knows that I is true, and can use the critical data knowing that it is consistent and up-to-date. The requirement for consistent critical data is easy to attain with monitors, and consequently monitors are a useful and understandable mechanism for dealing with concurrency.

This brief introduction to monitors is intended to give the general idea of how they are used. Details about monitors, as supported by the Concurrent Euclid language, are covered in later chapters, where it is shown how monitors are a convenient feature for structuring operating systems.

Monitors have been used as the basis of the SUE/11 operating system; this is a small special purpose system used to run PL/I subset (SP/k) jobs on a PDP-11 minicomputer. Compilers have been implemented that support the Pascal language augmented by monitors. The most notable extended versions of Pascal are called Concurrent Pascal, Modula and Pascal Plus. Interesting special purpose operating systems have been written in Concurrent Pascal and used on PDP-11 minicomputers. Tunis, which is an operating system that is compatible with Unix, has been written using monitors as supported by the CE language; this is described in a later chapter.

We have discussed mutual exclusion, the block/wakeup facility, and message passing as fundamental problems and constructs in concurrent programming. We will now consider another fundamental concurrency problem: deadlock.

THE DEADLOCK PROBLEM

In concurrent programming, a process sometimes must wait until a particular event occurs. If the event takes place and the waiting process is awakened, then there is no problem. But if the event never occurs, the process will be blocked forever! We say a process is *deadlocked* when it is waiting for an event that can never occur.

A simple example of deadlock can occur in a system with two processes, P1 and P2, and two resources, R1 and R2. Suppose process P1 acquires resource R1 and P2 acquires R2. Then P1 requests R2. Since R2 is already allocated, P1 is blocked until R2 becomes available; presumably R2 will eventually be released by P2. But now P2 requests resource R1. Since R1 is already allocated, P2 is blocked until R1 becomes available. The situation is illustrated in this diagram.

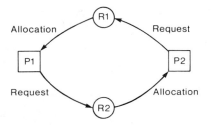

We are assuming that a resource cannot be pre-empted from a process - which means that once a resource is acquired by a process, it will not become available to another process until the acquiring process releases it. The result is that the cycle of waiting conditions illustrated in the diagram can never be satisfied. Processes P1 and P2 are deadlocked. They will remain blocked until special action is taken by an "external force" such as the operator or the operating system.

Unfortunately, not all deadlocks are as simple as this example. More complex cases can arise when there are many processes and many resources. The blocking that results in a deadlock can arise from any of the synchronization operations that allow processes to wait; for example, the semaphore P operation, the Lock operation, the Receive operation, and attempting to enter a monitor.

In operating systems, deadlocks can be expensive or disastrous. We will give two examples that are taken from production operating systems. The first example involves spooling, which means temporarily storing input and output records on disk, to provide buffering for devices such as card readers and printers. In spooling systems, deadlock can occur due to competition for disk space. The problem occurs when the space becomes completely filled with input records for jobs waiting to execute and output records for jobs not finished executing. If there is no way to recover the space from a partially executed job (and there is not in many systems), then the only way to recover from such a deadlock is to restart the system. A crude but usually effective solution to this problem is to prohibit the spooling of new jobs when too much spooling space is occupied, say, more than 80%.

The second example of deadlock can easily be caused by a hostile user, given that the system supports PL/I with multitasking (concurrent programming). The following four-line PL/I program does the trick.

```
REVENGE: PROCEDURE OPTIONS(MAIN,TASK);
    DECLARE(E)EVENT;
    WAIT(E);
    END;
```

This program does nothing but wait for an event that will never occur. The user will not be charged for CPU or I/O because the program uses neither. However, any resources allocated to this program, such as the memory it occupies, will remain idle until the deadlock is detected and removed, either by the operating system or by a keen-witted operator. The next section explains how certain types of deadlock can be automatically detected.

DETECTING DEADLOCK

In a multiprogramming system, users' jobs compete for the available resources. For example, two jobs may simultaneously need to use a tape drive. Resources such as tape drives are called *re-usable*, because after they have been used by one process, they can be re-used by another process. In this section we will show how deadlock can be detected when processes share re-usable resources. Each re-usable resource has the following properties:

There is a fixed total number of identical units of the resource. Each unit of the resource is either available (not allocated) or has been acquired by (allocated to) a particular process. A particular unit of a resource can be allocated to at most one process at a time. A process can release a unit of a resource only if the process has been allocated that unit. Units cannot be pre-empted; once a process has acquired a unit, the unit will not become available until released by the process.

The physical devices of the computer system, such as memory, tape drives and disks, can be thought of as re-usable resources. The number of units of some of these resources will depend on the allocation strategies of the computer system; for example, disks may be allocated in units of tracks, or cylinders or even entire disks. Certain information structures, such as files or linkage pointers for buffers, are re-usable resources. The process must request, acquire and release access to these information structures to guarantee that the structure can be inspected or updated without interference from other processes. Previously, we showed that mutually exclusive access to certain data was required in critical sections; we are now pointing out that such critical data is equivalent to a re-usable resource (with a single unit).

We can represent a system of processes and re-usable resources by a graph having nodes for each process and resource. The units of a resource are shown by small circles inside the resource nodes, as illustrated here:

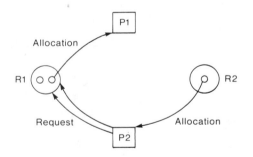

This graph shows a system with two processes and two resources. Resource R1 has two units and R2 has one unit. One of the units of R1 has been allocated to P1. P2 has acquired the unit of R2 and has requested both units of R1. P2 will be blocked until P1 releases its unit of R1.

As processes request, acquire and release units of resources, the graph changes. Suppose the system has two processes and a single resource with three units, one of which is allocated to process P2, as shown here.

Now process P1 requests two units of the resource:

Since two units are available, P1 acquires them:

Next process P1 releases one of the units:

In these graphs, an arrow is drawn from a process to a resource for each request of a unit and an arrow is drawn from a unit of a resource to a process for each allocated unit.

What we would like is a method of analyzing a graph to determine if there is a deadlock; that is, to see if some processes can never be granted their requests. It turns out that there is a relatively easy way to do this, using "graph reductions". We say a graph can be *reduced* by a process if all the process's requests for units can be granted. For example, in this graph we can reduce by process P1 because its request for a unit of R2 can be granted.

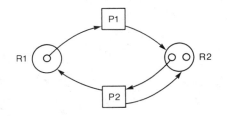

We reduce a graph by a process by deleting all arrows to or from the process. For example, when the above graph is reduced by P1, we get this graph:

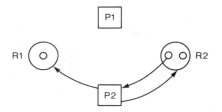

Since the requests by P2 can now be granted, we can further reduce the graph to the following:

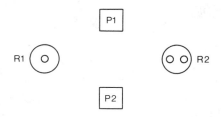

We say this graph is *completely reduced* because there are no more arrows and thus no more allocations or requests.

Essentially, the reductions determine whether processes can release their resources so other processes can receive their requests. The reductions will either delete all arrows, or they will leave certain processes not reduced. The processes that could not be reduced are those that are deadlocked. The following theorem can be proved:

There are no deadlocked processes if and only if the graph is completely reducible.

The order of reductions -- when different orders are possible -- is immaterial because the same final graph is obtained regardless of the order. The reason is that each reduction can allow new reductions (because new units are released), but can never prevent other reductions from taking place.

We can now give the algorithm for seeing if processes are deadlocked. The algorithm repeatedly checks to see if processes can be reduced, until none can be. Then it sees if all processes were reduced, i.e., if the graph was completely reduced. If all processes were reduced, there was no deadlock. Before the algorithm begins, each element of the Boolean vector called reduced is set to false, indicating that no processes are initially reduced.

```
   {Algorithm to detect deadlock}
 1 reducedProcesses := 0
 2 reduction := true
 3 {Repeat until no more reductions are possible}
 4 loop
 5     exit when not reduction
 6     reduction := false
 7     p := 1
 8     loop {Try to reduce each process}
 9         if not reduced(p) then
10             if process p can be reduced then
11                 reduced(p) := true
12                 reducedProcesses := reducedProcesses + 1
13                 reduction := true
14                 Reduce by process p
15             end if
16         end if
17         exit when p = numberOfProcesses
18         p := p + 1
19     end loop
20 end loop
21 completelyReduced := (reducedProcesses=numberOfProcesses)
```

Lines 10 and 14 of this algorithm require that we define a particular data structure to represent the graph. If there is only one resource, which has several units, the graph can be represented by the following variables:

req: a vector with a subscript range from 1 to numberOfProcesses, giving the present request by each process.

alloc: a vector with a subscript range from 1 to numberOfProcesses, giving the present allocation to each process.

avail: an integer giving the present number of available units.

Assuming these are appropriately initialized before our algorithm is executed, line 10 can be written this way:

if req(p) < = avail then

Line 14 can be written as:

avail := avail + alloc(p)

If there are several resources, each with several units, then req, alloc and avail must additionally be subscripted by a resource number. In this case we introduce the flag reduceByP and replace line 10 by:

```
reduceByP := true
r := 1
loop {For each resource}
    if req(p)(r) > avail(r) then
        reduceByP := false
    end if
    exit when r = numberOfResources
    r := r + 1
end loop
if reduceByP then
```

Note that req(p)(r) is an example of Euclid notation for double subscripting. Now line 14 is written as:

```
r := 1
loop {For each resource}
    avail(r) := avail(r) + alloc(p)(r)
    exit when r = numberOfResources
    r := r + 1
end loop
```

If the graph is represented by a linked list, different replacements for lines 10 and 14 would be required.

Our algorithm can be used each time a request is made that cannot be immediately granted. Alternatively, the algorithm might be invoked only when there is reason to suspect a deadlock, for example, when a process has been blocked for a long time. An algorithm much like this one was used in the TOPPS language processor to detect deadlocks and to remove deadlocks similar to those that occur with events in PL/I. In many systems it is not practical to automatically detect deadlocks in this way because not enough information is available about the interactions among processes.

CHAPTER 2 SUMMARY

In this chapter we have introduced programming language features that support the following basic concurrency requirements:

Concurrent execution (cobegin/end, fork/quit/join).

Mutual exclusion (mutexbegin/end, lock/unlock, semaphores, monitors).

Block/wakeup (events, semaphores, wait/signal in monitors).

Message passing (send/receive, mailboxes, pipes).

The following additional important terms were discussed in this chapter:

Disjoint processes (independent processes) - processes that have no shared data.

Busy waiting - continual testing and re-testing of a condition until it becomes true. Generally, busy waiting is an unacceptable waste of CPU time.

Test and set instruction - an instruction that both tests and changes a value in an indivisible action; allows simple implementation of mutual exclusion via busy waiting.

Synchronization primitives - simple operations, such as P and V for semaphores, that allow processes to synchronize their activities. These can be used to implement mutual exclusion and block/wakeup.

Reusable resources - devices, files, or data structures that can be used by one process, then re-used by another, and so on. A process must request, acquire and then release such resources.

Deadlock - the situation in which one or more processes are blocked waiting for something that can never occur. Processes can become deadlocked when competing for re-usable resources.

CHAPTER 2 BIBLIOGRAPHY

This chapter has referred to several operating systems, namely, Tunis, Unix [Ritchie and Thompson], RC4000 [Brinch Hansen 1970], Thoth [Cheriton], VENUS [Liskov], SUE/11 [Greenblatt and Holt], and SUE/360 [Sevcik et al.]. The articles on Unix and RC4000 are especially interesting.

The TOPPS language is specified in detail by Czarnik et al. Dijkstra's basic work [1968] introduced semaphores, mutual exclusion and techniques for concurrent programming. Dekker's algorithm as presented in this chapter is based on Peterson's ingeniously simple solution [1981]. Hoare's article [1974] on monitors contains their formal definition along with an implementation in terms of semaphores and several interesting examples of monitors. Cashin [1980] describes the trade offs in designing message passing schemes. Hoare's article [1978] defines the CSP notation and gives a number of examples of their use. Examples of Ada rendezvouses are given by Ichbiah et al [1979]. Concurrent Pascal and Modula are described by Brinch Hansen [1975] and Wirth [1977]. More on the theory of deadlock can be found in Holt's article [1972].

Brinch Hansen, P. The nucleus of a multiprogramming system. *Comm. ACM 13*,4 (April 1970), 238-241, 250.

Brinch Hansen, P. The programming language Concurrent Pascal. *IEEE Trans. on Software Engineering SE-1*,2 (June 1975), 199-207.

Cashin, P.M. Inter-process communication. Report 8005014, Bell-Northern Research, Ottawa, Canada, June, 1980.

Cheriton, D.R., Malcolm, M.A., Melen, L.S., Sager, G.R. Thoth, a portable real-time operating system. *Comm. ACM* 22,2 (February 1979), 105-115.

Czarnik, B. (editor), Tsichritzis, D., Ballard, A.J., Dryer, M., Holt, R.C., and Weissman, L. A student project for an operating systems course. CSRG-29, Computer Systems Research Group, University of Toronto (1973).

Dijkstra, E.W. Cooperating sequential processes. In *Programming Languages.* (F. Genuys, editor), Academic Press (1968).

Greenblatt, I.E. and Holt, R.C. The SUE/11 operating system. *INFOR*, Canadian Journal of Operational Research and Information Processing 14,3 (October 1976), 227-232.

Hoare, C.A.R. Monitors: an operating system structuring concept. *Comm. ACM 17*,10 (October 1974), 549-557.

Hoare, C.A.R. Communicating sequential processes. Comm. ACM 21,8 (Aug. 1978), 666-677.

Holt, R.C. Some deadlock properties of computer systems. *Computing Surveys 4*,3 (September 1972), 179-196.

Ichbiah, J.D., Barnes, J.G.P., Heliard, J.C., Krieg-Breuckner, B., Roubine, O., Wichman, B.A. Rationale for the design of the Ada programming language. ACM SIGPLAN Notices 14,6 (June 1979).

Liskov, B.H. The design of the VENUS operating system. *Comm. ACM 15*,3 (March 1972), 144-149.

Peterson, G.L. Myths about the mutual exclusion problem, Information Processing Letters 12,3 (June 81), 115-116.

Ritchie, D.M. and Thompson, K. The Unix time-sharing system. *Comm. ACM 17*,7 (July 1974), 365-375.

Sevcik, K.C., Atwood, J.W., Clark, B.L., Grushcow, M.S., Holt, R.C., Horning, J.J., Tsichritzis, D. Project SUE as a learning experience. *Proc. FJCC 1972, Vol. 39*, 331-337.

Wirth, N. MODULA: a language for modular programming. *Software Practice and Experience Vol. 7*,1 (January-February 1977), 3-35.

CHAPTER 2 EXERCISES

1. Generalize Dekker's algorithm to handle n processes.

2. One method of implementing semaphores has the P operation decrement the semaphore count before testing the count's value. The result is that the count is sometimes negative, and the absolute value of the negative count gives the number of processes waiting. Give such an implementation of P and V in terms of mutexbegin/end and block/wakeup.

3. The parameterized version of mutexbegin/end is very similar to lock/unlock. The difference is that mutexbegin and mutexend are syntactically (statically) balanced brackets. Give the advantages and disadvantages of parameterized mutexbegin/end versus lock/unlock.

4. The parameterized version of mutexbegin/end allows separate mutual exclusion for separate collections of critical data. Show how, in certain cases, nesting of these constructs can lead to deadlock. Give a rule which a programmer can follow to prevent such deadlocks.

5. Implementing mutual exclusion by busy waiting can be very tricky. Is the following implementation correct? Explain.

```
mutexbegin:
    need(me) := true
    loop
        exit when not need(other)
    end loop


mutexend:
    need(me) := false
```

6. Implementing mutual exclusion by busy waiting can be very tricky. Is the following implementation correct? Explain.

```
mutexbegin:
    need(me) := true
    loop
        exit when not need(other)
        need(me) := false
        loop
            exit when not need(other)
        end loop
        need(me) := true
    end loop


mutexend:
    need(me) := false
```

7. The "banker's algorithm" is a method of preventing deadlock due to competition for re-usable resources. The method assumes that each process has a "claim" on resources, specifying its maximum need for resources. For example, process P1 might need at most 2 units of resource R1 and 5 units of resource R2. The operating system (the banker) receives requests from processes, and temporarily blocks any requests that could lead to deadlock. The operating system determines if a request is safe (cannot lead to deadlock) by seeing if an immediate request by all processes for the remainder of their claims leads to deadlock. Show how the deadlock detection algorithm given in this chapter can be used to see if requests are safe.

8. In some operating systems, notably the RC4000 system, there is exactly one mailbox per process. This means that messages are sent to processes rather than to mailboxes, because mailboxes have no separate identity. Give the advantages and disadvantages of separating mailboxes from processes.

9. High-level concurrent programming features can be based on the idea of resources. Essentially, a resource is anything that a process can wait for. "Reusable resources" were described in this chapter, and they can be used for mutual exclusion, as well as for allocation of physical devices. "Consumable resources" are like mailboxes; a release (send) operation to a consumable resource gives it another unit (a message) and a request (receive) operation retrieves a unit. Implement request/release for re-usable and consumable resources, using mutexbegin/end and block/wakeup.

10. In some systems deadlock can be prevented by pre-arranged conventions. One of the simplest and most effective of these is based on ordered (or hierarchic) re-usable resources. Each resource is in one of the classes 1, 2, up to n. A process must always request resources in order, from class 1 up to class n. Putting this another way, a process that holds a resource from class i is not allowed to request a resource from the same class or a lower numbered class. Prove that this convention prevents deadlock.

11. The deadlock detection algorithm in this chapter requires maximum time proportional to m times n^2, where m is the number of resources and n is the number of processes. Develop an algorithm that requires time proportional to m times n. (Hint: for each resource use a queue of processes ordered by request size.)

12. In a system with one resource type, find an algorithm to detect deadlock whose execution time is independent of the number of processes. Do not restrict requests to be for only one unit.

13. Explain why non-parameterized mutexbegin/end cannot be nested. Note that parameterized mutexbegin/end can be nested, given different parameters. Give a rule the compiler can enforce that prevents deadlock due to nested parameterized mutexbegin/end. Here is an example that can lead to such a deadlock.

```
P1: ...                          P2: ...
    mutexbegin (a)                   mutexbegin (b)
        mutexbegin (b)                   mutexbegin (a)
            ...                              ...
        mutexend (b)                     mutexend (a)
    mutexend (a)                     mutexend (b)
```

14. High in the Andes Mountains, there are two circular railroad lines. As shown in the diagram, one line is in Peru, the other in Bolivia. They share a section of track, where the lines cross a mountain pass that lies on the international border.

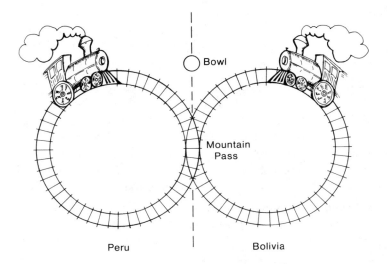

Unfortunately, the Peruvian and Bolivian trains occasionally collide when simultaneously entering the critical section of track (the mountain pass). The trouble is, alas, that the drivers of the two trains are blind and deaf, so they can neither see nor hear each other.

The two drivers agreed on the following method of preventing collisions. They set up a large bowl at the entrance to the pass. Before entering the pass, a driver must stop his train, walk over to the bowl, and reach into it to see if it contains a pebble. If the bowl is empty, the driver finds a pebble and drops it in the bowl, indicating that his train is entering the pass; once his train has cleared the pass, he must walk back to the bowl and remove his pebble, indicating that the pass is no longer being used. Finally he walks back to the train and continues down the line. If a driver arriving at the pass finds a pebble in the bowl, he leaves the pebble there; he repeatedly takes a siesta and re-checks the bowl until he finds it empty.

Then he drops a pebble in the bowl and drives his train into the pass. A smart aleck college graduate from the University at La Paz (Bolivia) claimed that subversive train schedules made up by Peruvian officials could block the Bolivian train forever. (Explain). The Bolivian driver just laughed and said that could not be true because it never happened. (Explain). Unfortunately, one day the two trains crashed. (Explain).

Following the crash, our college graduate was called in as a consultant to ensure that no more crashes would occur. He explained that the bowl was being used in the wrong way. The Bolivian driver must wait at the entry until the bowl is empty, drive through the pass and walk back to put a pebble in the bowl. The Peruvian driver must wait at the entry until the bowl contains a pebble, drive through the pass and walk back to remove the pebble from the bowl. Sure enough, his method prevented crashes. Prior to this arrangement, the Peruvian train ran twice a day and the Bolivian train ran once a day. The Peruvians were very unhappy with the new arrangement. (Why?)

Our college graduate was called in again and was told to prevent crashes while avoiding the problem of his previous method. He suggested that two bowls be used, one for each driver. When a driver reaches the entry, he first drops a pebble in his bowl, then checks the other bowl to see if it is empty. If so, he drives his train through the pass, stops it and walks back to remove his pebble. But if he finds a pebble in the other bowl he goes back to his bowl and removes his pebble. Then he takes a siesta, again drops a pebble in his bowl and re-checks the other bowl, and so on, until he finds the other bowl empty. This method worked fine until late in May, when the two trains were simultaneously blocked at the entry for many siestas. (Explain.)

Chapter 3

CONCURRENT EUCLID: SEQUENTIAL FEATURES

Chapter 2 discussed various notations for handling concurrency, including monitors. The rest of this book uses the Concurrent Euclid (CE) language, which supports monitors.

CE is an important example of a high-level language that is well-suited for writing operating systems, including basic software such as device management. CE is a higher level language than Unix's C language, which does not support strong type checking or concurrency. Given a language like CE, we do not need to resort to the complexity and unreliability of assembly language except for highly specialized purposes, such as saving registers following an interrupt.

This chapter introduces the sequential features of CE and the next chapter gives its concurrency features. The appendix "Specification of Concurrent Euclid" defines the language in detail.

HISTORY

Euclid was designed in 1976 as a language for developing verifiable systems software, i.e., system software that can be proven correct. In a two year effort beginning in 1977 a subset called Toronto Euclid was implemented jointly by the University of Toronto and I.P. Sharp Associates with the support of the USA Department of Defense and the Canadian Department of National Defense. Toronto Euclid was used experimentally at the University of Toronto for implementing compilers and for developing the

Tunis operating system, which is compatible with Unix. This research lead to the design and implementation of the language Concurrent Euclid in 1980-81. Concurrent Euclid omits the complex features of Euclid and adds features not in Toronto Euclid that were found necessary for systems programming. The most notable additions were concurrency and separate compilation.

GOALS OF CONCURRENT EUCLID

Concurrent Euclid inherits constructs from Euclid that were designed with verification in mind. Generally these constructs not only aid verification; they also enhance reliability and understandability. These constructs sometimes make a program harder to write. This is because the construct requires the programmer to document his program more thoroughly; for example with each procedure he must give a list of variables accessed by the procedure. It is sometimes not so easy to get the CE compiler to accept a program, because CE imposes many restrictions on the program such as strong type checking. The philosophy of Euclid is that the compiler should detect as many errors as possible; this is done by disallowing dangerous or unlikely constructs. Thus the compiler helps increase reliability and decrease maintenance by locating bugs at compile time. This is better than leaving these bugs to be located by the relatively costly and undependable process of testing. It is vastly superior to leaving these bugs in a software product that is delivered to a user community.

Since CE is intended for writing system software, it provides "escape" features to override compile-time checks; for example, "converters" can be used to explicitly defeat strong type checking. The programmer uses these escape features at his own risk, presumably only when they are actually needed.

CE is designed to allow efficient generated code and small, fast, highly portable compilers. The CE compiler implemented at the University of Toronto consists of four passes and runs in about 50K bytes. The last pass is a replaceable code generator with versions that generate high quality code for various computers including the PDP-11, VAX, MC68000 and MC6809.

In summary, CE was designed to support implementation of highly reliable, high performance software, such as operating systems, compilers and embedded microprocessor software.

COMPARISON WITH PASCAL

Euclid is based on Pascal and borrows Pascal's elegant data structures. Various features of Pascal were "purified" to allow easier verification; for example in Euclid and CE, functions are prevented (by the compiler) from having side effects. CE can be thought of as a cleaned up version of Pascal that adds features needed for systems programming. The major features CE adds to Pascal are:

(1) Separate compilation. Procedures, functions and modules can be separately compiled and later linked together. Under Unix, these use the standard linker (ld) and are easily interfaced to programs written in languages like C and assembler.

(2) Modules. A module is a syntactic packaging of data together with procedures/functions that access the data.

(3) Concurrency. Monitors and processes are supported. There is a signal statement and a wait statement. A "busy" statement allows CE to be used as a simulation language.

(4) Control of scope. Names of variables, types, etc. are not automatically inherited by scopes. Import and export lists are used to control the scope of names.

(5) Systems programming constructs. These include variables at absolute addresses; such variables can be device registers in computers with memory mapped input/output.

There are some Pascal features, such as enumerated types, that CE does not support. CE does not allow procedures and functions to be nested inside procedures and functions.

BASIC DATA TYPES

CE has the traditional basic data types of Pascal, except float and enumerated types. There are several ranges of integers to reflect hardware data types. These basic types are:

Name	Values	Allocation
ShortInt	0..255	byte
SignedInt	-32768..32767	16-bit
UnsignedInt	0..65535	16-bit
LongInt	signed integer	32-bit
Boolean	false..true	byte
Char	a character	byte
AddressType	integer	address size
Pointer	address	address size

Besides these, there are subranges, for example, the type 1..10, which is an integer subrange. The values and allocations are those used by the CE compilers for the PDP-11, VAX, Motorola 68000 and Motorola 6809. Implementations for other computers provide at least as much range as this table specifies for ShortInt, SignedInt and UnsignedInt. Other implementations are not restricted to use the same space allocation for these basic types.

STRUCTURED DATA TYPES

CE inherits the structured types of Pascal, namely (1) arrays, (2) records and (3) sets. The following are example declarations using these types.

(1) Arrays. These are vectors of elements.

        var a: array 1..10 of SignedInt
        var str: packed array 1..5 of Char := 'Hello'
        var matrix: array 1..5 of array 1..5 of LongInt

Variable a is an array of 10 SignedInt elements. Variable str is a "string". CE inherits from Pascal the definition that a "packed" array of characters with lower bound of 1 is considered to be a string.

A quoted value, such as 'Hello', is a value of type string. As can be seen in this example, CE allows a variable to be initialized in its declaration.

CE does not provide multi-dimensioned arrays, but it allows arrays of arrays, which are equivalent. For example, the variable declared as matrix has five "rows" each of which has five LongInt elements. Row three of the matrix is written as "matrix(3)". Note that CE uses parentheses, and not square brackets, around subscripts. Element five of row three is written as matrix(3)(5). It is not legal to write matrix(3,5).

(2) Records. These are equivalent to Pascal records and PL/I's structures.

```
var r:
    record
        var status: Boolean
        var count: SignedInt
    end record
```

This example declares r to be a record with fields called status and count. A field is accessed using a dot, so "r.count:=2" assigns two to the count field. Fields of records cannot be initialized as a part of the record's declaration.

(3) Sets. These are essentially bit strings.

```
var s: set of 0..2
```

Set variable s is implemented on a PDP-11 as bits number 0, 1 and 2 in a byte. These bits can be individually changed and inspected.

LITERAL VALUES

We call an item such as 211 a *literal* because it denotes its own value. Other languages use the term "constant" to mean what we will call a literal. We will use the term *constant* in a different sense, namely to denote a value that cannot change in a particular scope. We now give examples of literals.

(1) Integer literals. There are three kinds:

```
decimal, e.g., 921 and 483649215
octal, e.g., 11#8
hexadecimal, e.g., 0E4F#16
```

A hexadecimal literal must begin with a decimal digit, so we write 0E4F#16 instead of E4F#16.

(2) Booleans. The literals are true and false.

(3) Characters. A character literal is a dollar sign followed by the character's value. For example, this declaration initializes variable c to the

value X.

> var c: Char := $X

Similarly, $a, $q and $? represent small a, small q and question mark. Peculiar characters require double dollar signs, for example,

> $$S means a space (a blank)
> $$N means a new line character
> $$T means a tab character
> $$E means an end-of-file character (octal zero in ASCII)
> $$$ means a dollar sign

(4) Strings. A string literal consists of characters inside single quote marks. Here are two string literals.

> 'Here is a new line $N and an EOF $E'
> 'And here is a quote $' too'

Any of the peculiar characters such as end-of-file characters are written inside a string using a dollar sign. If an actual dollar sign is wanted, it is written as two dollar signs.

(5) Sets. Properly speaking, sets do not have literals. Instead they have "set constructors", which are analogous to literals.

> type S = set of 0..2
> var x: S
> x := S(0,2)

Variable x's value has three bits numbered 0, 1 and 2. S(0,2) is a set constructor. The assignment gives x a value with bits 0 and 2 on, but with bit 1 off.

THE I/O PACKAGE

Since CE is a system language, it is adept at implementing basic support such as I/O. The CE language does not define input/output operations, but relies on the programmer to implement these. A standard I/O package has been defined (and programmed in CE) to support convenient, portable I/O.

To use this I/O package, its definition must be copied into the program at hand. To include this definition as a part of a program, one writes:

include '%IO1'

More features of the I/O package are available by changing IO1 to IO2, IO3 or IO4. See the appendices for details. The I/O package includes procedures for reading and writing characters, integers and strings. (Note: the percent sign in '%IO1' tells the compiler to search a particular library to find the file called IO1.)

For example, here is a statement that writes a string and one that writes a character.

IO.PutString('HiNE')
IO.PutChar($H)

The first prints the string "Hi" and starts a new line as denoted by $N. The final $E is required by PutString to mark the end of the string to be written. The second statement prints the character H.

The input/output package is implemented as a CE module named IO that contains the procedures PutString, GetString, etc. We write IO.PutString to mean: use the procedure PutString in the module named IO.

We can read character variable ch as follows:

IO.GetChar(ch)

Under an operating system like Unix, the Get and Put operations by default read from and write to the user's terminal.

Integer i can be read and written as follows:

IO.GetInt(i)
IO.PutInt(i,4)

After reading i, this prints i right justified in a field of width 4. If the width specified in PutInt is not sufficient to print i's value, the field is automatically widened. So the statement IO.PutInt(i,1) writes only as many characters as it takes to give i's value.

The complete I/O package provides a rich set of procedures; under an operating system such as Unix these support all the system's file operations, including open, close, seek, accessing multiple files, etc. On a bare machine, say an unadorned MC6809, the I/O package is implemented by procedures that access device registers to directly implement physical input/output.

A COMPLETE CE PROGRAM

Enough features of CE have been introduced so that we can give an example program that actually accomplishes something. The following example reads and prints a file. We show numbers in the left margin of the program so we can refer to parts of the program, but in an actual program these numbers would not be present.

```
1       var Example:
2           module

3               include '%IO1'

4               { Print characters up to a period }

5               initially
6                   imports(var IO)
7                   begin

8                       var ch: Char
9                       IO.PutString('Test starts$N$E')
10                      loop
11                          IO.GetChar(ch)
12                          IO.PutChar(ch)
13                          exit when ch = $.
14                      end loop

15                  end {of initially}

16          end module
```

The name of our program is "Example" as given in line 1. Lines 1, 2 and 16 are useless and are analogous to the useless "program" header in Pascal and the useless "procedure options(main)" in PL/I. These lines are required in CE so that a program has the same format as a "module". Line 3 allows us to use the I/O package. Line 4 is a comment. In general a comment is any character sequence enclosed in braces, just like the Pascal convention. However, Pascal's convention of (* and *) as alternatives for braces is not allowed.

Technically speaking our program executes by initializing the data in its module. That is why lines 5 through 7 and 15 enclose the logic of our program. In this example there is no data to initialize, and the "initially" procedure simply executes the logic of our program.

Line 6 states that the "initially" procedure is going to use the IO module. If line 6 were omitted, the compiler would consider the calls to Get and Put procedures in lines 11 and 12 to be illegal. The IO module is imported as a variable (var) meaning that it is going to be changed. It may come as a surprise that the IO module can be changed, but technically this is true because it "contains" the input/output files and those are modified by Puts.

Finally we are down to the meat of the program, which is lines 8-14. Line 8 declares character variable ch which is read in line 11 and written in line 12. Line 9 prints "Test starts" and begins a new line.

Lines 10 and 14 are the loop construct that specifies that the enclosed statements are to be repeatedly executed. This looping will continue until an exit statement (or return statement) terminates the loop. Statement 13 terminates the loop when ch is a period. A loop may contain several exit statements and these exits may be nested inside other statements. For example, we could insert the following if statement just before "end loop" to handle a premature end-of-file.

```
if ch = $$E then
    IO.PutString('$NMissing period$N$E')
    exit
end if
```

Note that the exit statement inside the if statement does not use a "when" clause.

OTHER CONTROL CONSTRUCTS

We have seen examples of the use of the loop statement and the if statement. The only other control constructs are the else and elseif clauses of an if statement and the case statement. The use of else and elseif is illustrated by the following loop that counts digits, white space (blanks, tabs and new-line characters), and other characters up to an end-of-file.

```
var digits: SignedInt := 0
var whites: SignedInt := 0
```

```
var others: SignedInt := 0
var ch: char
loop
    IO.GetChar(ch)
    exit when ch = $$E {Stop on end-of-file}
    if ch >= $0 and ch <= $9 then
        digits := digits + 1
    elseif ch = $$S or ch = $$T or ch = $$N then
        whites := whites + 1 {Spaces, tabs and new lines}
    else
        others := others + 1
    end if
end loop
```

An if statement can contain any number of elseif clauses, followed by an optional else clause; these are followed by "end if". The tests following if and elseif are evaluated in sequence until one is found to be true; the statements in the clause corresponding to the true test are executed and then control skips to just beyond "end if". If none of the tests are true, then the else clause if present is executed.

Using the same variables as above, we now re-write the loop using a case statement.

```
loop
    IO.GetChar(ch)
    exit when ch = $$E {Stop on end-of-file}
    case ch of
        $0, $1, $2, $3, $4, $5, $6, $7, $8, $9 =>
            digits := digits + 1
            end $0
        $$S, $$T, $$N =>
            whites := whites + 1
            end $$S
        otherwise =>
            others := others + 1
    end case
end loop
```

The case statement is much like an if statement containing elseif's, except that it is generally faster when there are several values to test. On the other hand, the case statement may be larger, because it is typically implemented using a run-time table containing as many addresses as the range of the case labels. (The "otherwise" label does not affect the table size.) The otherwise clause of a case statement is optional. If it is omitted then the selector (ch) is required to match one of the labels.

RUNNING UNDER UNIX

CE can be compiled and run under the Unix operating system using conventions similar to those for other languages such as C. Suppose that the example program of the last section is in a Unix file named "example.e" and that a file named "data" contains the line:

> File contents.

The following Unix session might take place:

> % cec example.e
> % example.out < data
> Test starts
> File contents.
> %

In the first line, Unix types its prompt character %. The user types "cec" (Concurrent Euclid compiler) and the name of the file to be compiled (example.e). If the compiler finds errors in the program, it prints appropriate messages on the terminal. But if the program is cleanly compiled, as is the case here, the compiler remains silent. The compiler produces an executable version of the program and puts it into a file named example.out.

Following the next prompt character, the user types "example.out < data" meaning that Unix should execute the program, using the contents of the file named "data" as input. When the program executes it prints "Test starts", followed by a new line. Next it reads and prints characters up to a period, thus producing the output "File contents." Unix again prints its prompt character and waits for the user's next command.

CE uses a suffix naming convention to distinguish various kinds of files. For example, the ".e" suffix signifies a source CE file. The various suffixes for the example program are now illustrated.

example.e	source CE program
example.s	assembly language version
example.o	linkable object version (object module)
example.out	executable object version (load module)

These assembly and object versions are for a PDP-11 or VAX. There are other forms for different target machines, for example:

example.t	assembly for MC68000
example.sout	executable object for MC68000
example.u	assembly for MC6809
example.tout	executable object for MC6809

The various options for using the compiler to produce such forms can be found out under Unix by typing "cec -help".

A SIMPLE PROCEDURE

CE has procedures that are essentially the same as those in Pascal. Here is one that increments its argument.

```
procedure Inc(var i: SignedInt) =
    begin
        i := i + 1
    end Inc
```

If variable k has the value 4 then the statement "Inc(k)" sets k to 5. The "var" in the declaration of i specifies that i can be changed, and that any change to i changes its corresponding actual parameter (k in this example). If this "var" were omitted, the compiler would consider the assignment "i:=i+1" to be illegal. This is different from Pascal in which this assignment would still be legal, and would change i but not k. In CE, a parameter declared without var is called a "constant" parameter, meaning that within its scope (down to "end Inc") its value will not change.

NESTING OF CONSTRUCTS

Generally, nested constructs in CE are bracketed at their start by the construct name and terminally by "end" followed by the construct name, for example:

record	...	end record
module	...	end module
loop	...	end loop
if	...	end if
case	...	end case

The exception to this rule is "begin". When "begin" is an executable statement, it ends with just "end". Similarly, the "begin" of an "initially" procedure is terminated with just "end". But when begin-end is used to bracket the body of a named procedure (or function), such as the "Inc" procedure, then the name of the procedure must follow "end", as in "end Inc".

AN EXAMPLE MODULE

One of the most important language constructs of CE is the "module", which is used to package data together with procedures and functions that use the data. A typical example of this is the implementation of a stack where the data consists of variables representing the stack's top and contents.

A stack of integers with maximum depth 10 can be programmed in CE this way:

```
1       var Stack:
2           module
3               exports(Push,Pop)

4               const depth := 10
5               var top: 0..depth
6               var contents: array 1..depth of SignedInt

7               procedure Push(i: SignedInt) =
8                   imports(var top, var contents)
9                   begin
10                      top := top + 1
11                      contents(top) := i
12                  end Push

13              procedure Pop(var i: SignedInt) =
14                  imports(var top, contents)
15                  begin
16                      i := contents(top)
17                      top := top - 1
18                  end Pop

19              initially
20                  imports(var top)
21                  begin
```

```
22                      top := 0
23              end

24          end module
```

Again we have numbered the lines of this example so we can refer to them easily, but the actual CE program does not have these numbers.

The first thing to notice is that this module has the same form as a main program, because a main program is actually a module.

Lines 1, 2 and 24 serve to bracket the construct and give it the name "Stack". Line 3 states that outside of these brackets, the only observable parts of the stack are its procedures Push and Pop. For example, outside the module we can use the statement "Stack.Push(6)" to put value 6 onto the stack.

Lines 4-6 declare the module's data. The value "depth" is a const (constant). It is called a *manifest constant* because its value is known at compile time. Manifest constants (and manifest expressions) can be used to declare the ranges of integers and of array subscripts as is done on lines 5 and 6.

Lines 19-23 are the initializing procedure for the module; they set top to zero. In a production program this initialization procedure would probably be omitted, because it is easier to re-write line 5 as:

```
var top: 0..depth := 0
```

However, we have used the "initially" procedure in this example to illustrate the use of this construct.

Lines 7-18 contain the Push and Pop procedures. Push has a non-var (constant) parameter, while Pop requires a var parameter to return the stack's top value. The Push procedure imports top and contents as variables (as var) because it modifies both. Without this import list, the compiler would disallow the assignments on lines 10 and 11. The import list of Pop is similar, but contents is imported non-var, because it is not modified.

We can put our Stack module to use to read a list of positive numbers (at most 10 of them) and print them in reverse order.

```
{Read integers up to a zero and print in reverse order}
var Reverse:
    module
        include '%IO1'
```

```
            include 'stack.e'
            initially
                imports(var IO, var Stack)
                begin
                    var x: SignedInt
                    stack.Push(0)
                    loop
                        IO.GetInt(x)
                        exit when x = 0
                        Stack.Push(x)
                    end loop
                    loop
                        Stack.Pop(x)
                        exit when x = 0
                        IO.PutInt(x, 8)
                    end loop
                end
        end module
```

This program pushes a zero onto the stack and then pushes input integers onto the stack till it finds a zero. Then it pops and prints these values.

As shown here, we have assumed that the Stack module is in a file called "stack.e" and we have copied it into our program by an "include" statement. Alternatively, we could actually place the stack module in our program and not use an include statement.

Notice that our main program, called Reverse, is a module that contains another module, Stack. In general, modules can contain modules which can contain modules, etc. The "initially" procedures of modules in a program are executed in textual order from the top to the bottom of the program.

The initially procedure of our main program imports Stack as "var". This is required because Push and Pop are used and they change values representing the stack. Actually, the rule is that a module must be imported var wherever one of its procedures is to be invoked.

NAMING CONVENTIONS

Our example programs illustrates the style of capitalization that is encouraged in CE. Names of procedures, functions, converters, types, modules and monitors begin with capital letters. Names of constants and variables begin with small letters. A name, such as PutString, that consists of multiple words has the first letter of each word capitalized, except for the first word of constants and variables. (Note: this style is not enforced by the compiler; it considers capital and small letters in identifiers to be equivalent.)

Names should be chosen to be meaningful; a name like BufferManager is far better then Bfmg. CE names can be long (up to 50 characters) so it is relatively easy to invent good ones. Unfortunately, externally known names, in particular the names of external procedures, functions and modules (see below), are truncated by some operating system utilities. The Unix (version 7) linker uses only the first 7 characters of these external names, so we must use care when inventing these names.

RUNNING ON A BARE MICROPROCESSOR

As has been mentioned, the I/O package for CE can be implemented to operate on a bare microprocessor. The simplest version of such a package does synchronous input/output; that is, it does a "busy wait" wasting the processor's power whenever input/output is in progress. Here is a partial implementation for running on an MC6809 microprocessor.

```
var IO:
    module
        exports(PutChar, ... etc ...)

        var ttyData (at 9001#16): Char
        var ttyStatus (at 9000#16): set of 0..7
        pervasive const outputReady := 1

        procedure PutChar(c: Char) =
            imports(var ttyData, ttyStatus)
            begin
                loop
                    exit when outputReady in ttyStatus
                end loop
                ttyData := c
            end PutChar

        ... other procedures ...

    end module
```

We can use this module to print a character; for example, "IO.PutChar($H)" prints "H".

The data declared in the module consists of device registers. At hexadecimal locations 9001 and 9000 on the microprocessor's bus are the data and status registers for the terminal. When the terminal hardware is ready to receive another character, it turns on bit number 7 in its status register. So our loop burns up processor time until this bit is turned on. Assigning a character to the hardware data register (ttyData) causes the character to be transmitted to the terminal.

The declaration of the constant outputReady specifies that it is "pervasive". This means that it can be used in inner scopes such as that of PutChar without being explicitly imported. CE allows constants and type names to be pervasive, but variables must always be explicitly imported.

In many uses of microprocessors (and miniprocessors and maxiprocessors) it is not reasonable to waste processor time while input/output is in progress. When running on a bare machine, this can be avoided by using special procedures called BeginIO, WaitIO and EndIO. But this will have to wait till we get to CE's concurrency features.

NON-MANIFEST ARRAY BOUNDS

In CE, arrays must have manifest size, meaning that their bounds must be computable at compile-time. But there is an exception. When an array is the formal parameter of a procedure or function then its upper bound can be given as the keyword "parameter". This is done here in the PutString procedure.

```
procedure PutString
    (str: packed array 1..parameter of Char) =
    imports(PutChar)

begin
    var i: SignedInt := 1
    loop
        exit when str(i) = $$E
        PutChar(str(i))
        i := i+1
    end loop
end PutString
```

This procedure can be used as a part of the IO module. It accepts a string, i.e., a value that is a packed array of characters with lower bound of one. It prints characters of the string until it encounters the end-of-file character ($$E). This character had better exist in the string or else this implementation of PutString is in trouble.

Notice that the PutChar procedure is imported because it is called by PutString. Procedures and functions are always imported non-var.

FUNCTIONS AND SIDE EFFECTS

In some programming languages there are operators that explicitly produce side effects in expressions, for example,

j := (i := i-1) + (i := 3*i)   { Not allowed in CE }

In this statement i is decremented by one and also multiplied by three. (In the C language, this would have "=" instead of ":=" to mean assignment.) Unfortunately, it is ambiguous whether the decrement or multiply is done first, and so we do not know what values j or i will be left with. Versions of C compilers for different machines produce different results. Obviously this is not a good state of affairs.

Here is another example of the danger of side effects:

j := i + f(x)

Suppose that function f changes the value of i. Apparently this expression should use the original value of i. Many compilers that do code optimization will compile this statement as if it is

j := f(x) + i

because this saves putting the value of i in a temporary location during the execution of f. This will use the modified value of i. Unfortunately the resulting value of j is ambiguous. In languages like Pascal, C, Fortran and PL/I, different compilers may produce different results.

The decision in CE was to ban side effects in expressions so verification and understanding would be easier, and so optimization would not change results. We will give an example of a function and then will explain how this ban is enforced.

```
function Max(a: SignedInt, b: SignedInt)
    returns m: SignedInt =
    begin
```

```
        if a > b then
            return (a)
        else
            return (b)
        end if
    end Max
```

This function returns the larger of a and b. As in PL/I, the function result is given in return statements.

A function must return by executing a return statement that gives a value, as in "return (b)". A procedure can return either by "falling off" its end or by executing "return" (without the parenthesized return value).

In CE the name of the returned value, "m" in this example, cannot be accessed. This is different from Pascal where the function's result is set by assigning to the function's name. The only use of m in CE is to give the formal specification of a function. For example, Max can be specified by the following relation between a, b and m:

$$m >= a \text{ and } m >= b \text{ and } (m = a \text{ or } m = b)$$

In other words, $m = \max(a,b)$. Such a specification can be used in proving the function correct.

Now back to side effects. CE has no operators that cause side effects inside expressions. So functions are the only possible source of side effects. The following constraints are imposed on functions to prevent side effects in them. They are not allowed to have var parameters or to import anything var, so they cannot directly cause side effects external to themselves. Neither are they allowed to import procedures that import anything var, or to import procedures that import procedures that import anything var, etc. The compiler checks to prevent these things.

These constraints make functions behave beautifully, like honest mathematical functions! But there is some penalty for this beauty; for example in side-effect languages like C, one often sees constructs similar to:

```
    loop
        exit when GetC = $.    {Not legal in CE!}
    end loop
```

This loop skips characters till it reads a period. The GetC function has the side effect of changing the input stream (by reading a character). It is impossible to write GetC in CE, because it would need to access the IO package "var", which would be prevented by the compiler. In CE, GetC must be programmed as a procedure (not a function) that reads a character, as is done in IO.GetChar(c). The value of c must be tested in a separate statement. This seems a relatively minor inconvenience, which is incurred to avoid the danger of lurking bugs due to side effects.

POINTERS AND COLLECTIONS

CE has pointers that are implemented as addresses, similar to pointers in C, Pascal and PL/I. However, the pointers are safer in CE than in Pascal, safer in Pascal than in PL/I and safer in PL/I than in C. The danger of pointers is that since they are represented as machine addresses, an assignment to something pointed to may accidently destroy any variable (or code).

CE avoids some of this danger by defining "collections" as array-like objects, whose elements are located by pointers. For example if c is a collection and p is a pointer then c(p) accesses an element of c. We now give an example of a collection.

```
type R =
    record
        var kind: ShortInt
        var id: packed array 1..3 of Char
    end record

var c: collection of R
var p: ^c
...
c.New(p)      {Allocates element of c, i.e., a record}
c(p).id := 'ABC' {Written p^.id := 'ABC' in Pascal}
...
c.Free(p)     {De-allocates element of c}
```

We declare c as a collection of records with fields called kind and id. This declaration does not allocate any space for c or its elements, so c starts out as an empty collection. The declaration of p specifies that p can subscript c, i.e., that p can point to elements of c. The "New" statement is used to allocate elements of a collection; in this example, space is found for record R and the address of this space is placed in p. If space cannot be found, p is set to the collection's null value, which is written "c.nil". When an element of a collection is no longer needed, its space can be de-allocated by the

"Free" statement.

Pointers and allocations in CE behave like the analogous constructs in Pascal, and space for all collections can come from the same storage "heap". The difference is that in CE, it is always obvious that a pointer locates a value not just in the heap, but in a particular collection. In terms of correctness proofs this means that pointers behave like subscripts (except for problems of allocating and de-allocating elements). So, the proof techniques used for arrays can be applied to pointers (actually, to collections).

There is another advantage of collections. Since an access to an array element and a collection element has the same syntax, we can write algorithms without being concerned about whether we are using arrays or collections. This similarity is sometimes called *uniform referents* to data objects. Array subscripting is relatively safe because runtime checks can keep arrays in bounds; so the programmer may choose to test his algorithm using arrays. Later if he needs somewhat faster execution, he may change his declarations and allocations and use the unchanged algorithm with pointers.

CE requires that all identifiers be declared textually preceding their use. This can cause problems in a linked list that has elements that point to each other. It would seem that it is impossible to declare such a list because its declaration is circular (it references itself). We will use an example to show how CE solves this circularity problem.

```
type Element = forward
var list: collection of Element
type Element =
    record
        var id: Char
        var next: ^list
    end record
```

Our list has elements, each containing a subscript (pointer) to another element. We need to declare "list" using the type "Element", but we also need to declare "Element" using the collection "list". To break this vicious circle, CE allows a type to be declared as *forward*. Between the time the type is declared as forward and its actual declaration appears, it can be used only as the element type of a collection.

ALIASING AND THE BIND STATEMENT

When a programmer sees the statements

    x := 1
    y := 2

he expects that x ends up as 1 and y as 2. It only makes sense. But in most programming languages, such as Pascal and PL/I, this is not guaranteed to be true. The following Pascal fragment illustrates this unpleasant possibility.

```
{This is Pascal, not CE}
var x: integer;
procedure P(var y: integer);
    begin
        x := 1;
        y := 2     {Now x becomes 2!}
    end;
...
P(x)
```

In this example, variable x is passed to var parameter y, but procedure P also directly accesses variable x as a global variable. Since P has accessed a particular variable (shall we call it x or shall we call it y?) by more than one name, we say there is *aliasing*.

The trouble with aliasing is that things change (in this example, x changes) in ways that surprise us. This kind of nasty surprise is a source of elusive bugs. Besides that, aliasing makes it extremely difficult to formally define what a statement such as y:=2 means.

To avoid these problems, CE bans aliasing. This is done by placing constraints on those constructs of the language where a variable (or part of a variable) can be renamed. In CE there are only two such constructs, namely reference parameters and the "bind" statement. Before explaining these constraints we need to discuss the bind statement.

We can re-name a variable (or a component of a variable) by doing a bind, for example:

    bind var x to v(i)

This is a declaration that specifies that x is a new name for element i of array v. (This construct is similar to Pascal's "with" statement, but gives a

name to the variable or component.) The bind statement is (usually) implemented by computing the address of the component $v(i)$, assigning this to
x and then treating x as a pointer. The "var" in the bind statement is
optional and allows assignments to x. If $v(i)$ is accessed often in a group of
statements, binding to it typically results in smaller faster code because the
subscripting is done only once, in the bind.

To prevent aliasing, CE causes v to disappear until the end of the
current scope. Since it is sometimes desirable to bind to more than one
element of an array, CE allows multiple binds, as in

   bind(var x to $v(i)$, var y to $v(j)$)

If i and j are the same value then we are again in trouble, because x and y
will be aliases. This aliasing cannot in general be prevented by the compiler, because the values of i and j are not known until run-time. However,
the CE compiler does issue a warning (and some day may optionally emit
code to check that i and j are different).

We have seen how aliasing due to the bind statement is avoided.
Similar techniques are used to keep parameters from causing aliasing. The
call to the procedure named P in the Pascal example would not be allowed
in CE because the compiler would detect that P both imports and receives
as a var parameter the same variable. There remains the difficulty, analogous to multiple binds, of several var parameters representing the same
variable. For example, suppose procedure Q has var parameters x and y
and is called by the statement "$Q(v(i),v(j))$". If i and j are equal, then x
and y are aliases. As in the case of multiple binds, the CE compiler prints a
warning when it detects this possibility.

TYPE CONVERTERS

CE has strong type checking; this means that the compiler disallows
unlikely combinations of types such as adding the integer 14 to the Boolean
value "true" or multiplying a set value times a character. (See the
"Specification of Concurrent Euclid" in the appendix for details of the type
checking rules.)

However, there are circumstances where a less rigorous approach to
checking is appropriate. This is the case when a tightly packed data structure already exists and is to be manipulated by a CE program. We will take
as an example the analysis of the PDP-11 processor status (PS) word. The
PS is a 16-bit word on a PDP-11 giving certain information about the CPU.
The bits in the PS are as follows, where bit 0 is the lower order bit.

Bits 0-3:	Condition code, consisting of the 4 bits CVZN where: C=carry, V=overflow, Z=zero and N=negative. These are set according to the result of the instruction.
Bit 4:	Trace bit. If set, every instruction traps.
Bits 5-7:	Unused.
Bits 8-10:	Priority. This is a priority number in the range 0 to 7, that determines which hardware devices can interrupt the CPU.
Bits 11-15:	These bits specify which register set to use and the current and previous machine mode (kernel, supervisor or user mode).

It is not our purpose to give the details of the PS, but instead to show how it can be manipulated. Suppose we want to turn on the trace bit; we accomplish this as follows:

```
type WordSet = set of 0..15
const traceBit := 4
var PS (at 177776#8): WordSet
...
PS := PS +  WordSet(traceBit)
```

The "+" in the assignment statement means set union (also known as bit-wise inclusive "or"). This turns on bit 4 of the PS, which is located on the PDP-11 bus at octal location 177776. Inspecting or setting a particular bit in the PS is straight forward using the set operations "in", "not in", set union (+), set subtraction (-) and set intersection (*).

Suppose we want to print the value of the priority. This can be done by turning off the five high order bits (line 5 below) and shifting to the right (line 6).

```
1       converter WordSetToUnsigned(WordSet) returns UnsignedInt
2       var prtyBits: WordSet
3       const shift8 := 256  {2**8}
4       var prty: 0..7
5       prtyBits := PS - WordSet(11,12,13,14,15)
6       prty := WordSetToUnsigned(prtyBits) div shift8
7       IO.PutInt(prty,4)  {Print 0 to 7 in field 4 wide}
```

Line 1 declares WordSetToUnsigned to be a converter, which changes a WordSet to be an UnsignedInt. Line 3 defines the constant shift8 which we use to shift the priority eight bits to the right, to get a value in the range 0 to 7. Lines 5 assigns to prtyBits the value of PS with high order bits 11 to 15 turned off. Line 6 converts prtyBits to an integer, so that it becomes legal to divide it by shift8. This division moves the priority bits to the three lower order bits of the word, so it becomes an integer value in the range 0 to 7.

A converter, such as WordSetToInteger, allows us to violate the usual type checking rules of CE. A converter does not generate any code; it simply allows the bit pattern representing a value to be considered to be a value of another type. In analyzing the PS, we used a converter because the PS word contains both set-like and integer-like parts. Except in special situations like this, we should avoid converters, because their use is machine dependent and does not have the clean mathematical basis supported by strong type checking. In this example, we assumed that the types UnsignedInt and WordSet have the same size and alignment. This is true for CE on a PDP-11, but might not be true for other implementations of CE.

As another use of converters, consider the problem of inspecting and changing an arbitrary byte in memory. For historical reasons, these two operations are called Peek (inspect) and Poke (change). Poke is quite dangerous because it can wreck any byte that the hardware allows us to address.

Our implementation depends on the fact that CE defines AddressType, an integer subrange, to be the same size as a pointer. In the following, we define a converter from AddressType to a pointer. We define a collection called memory for the sole purpose of peeking and poking at arbitrary locations.

```
var memory: collection of ShortInt
type MemPtr = ˆ memory
converter AddrToPtr(AddressType) returns MemPtr

function Peek(location: AddressType)
    returns result: ShortInt =
    imports(memory, AddrToPtr)
    begin
        return(memory(AddrToPtr(location)))
    end Peek
```

```
procedure Poke(location: AddressType,
    value: ShortInt) =
    imports(var memory, AddrToPtr)
    begin
        memory(AddrToPtr(location)) := value
    end Poke
```

As can be seen, we never allocate or free the elements of the collection called memory. Instead, we convert the integer named location and treat it as a pointer. The pointer accesses what seems to be an element of a collection but is really the byte whose address is "location".

Peek and Poke work for architectures that supports byte addressing, but may not work for an architecture such as the PDP-10 which allows direct addressing of words but not of bytes. In general, one should use converters and routines such as Poke only when they are really needed, as they are dangerous and machine dependent.

SEPARATE COMPILATION

CE allows procedures, functions, modules and monitors to be compiled separately and later linked together. For example, the IO module is pre-compiled and is linked with most CE programs.

Suppose that the Max function has been previously compiled and its object code has been saved in a file. When we want to use it, we place the following *stub* (or interface) for Max in our program:

```
function Max(a: SignedInt, b: SignedInt)
    returns m: SignedInt =
    external
    ...
x := Max(i,0)  {use of the Max function}
```

This stub for Max is just like the actual Max function except that its body is replaced by the keyword external. The stub is required in any program that calls Max. It is used by the compiler to determine the types of the parameters and result of Max.

Separate compilation of procedures is done analogously to functions, namely, the stub for a procedure has the body replaced by the keyword external.

The stub for a separately compiled module (or monitor) is marked as "external" and omits the declarations of the module's variables and unexported procedures; also, the bodies of its entry points are replaced by the keyword external. For example, the Stack module, which was presented previously, would have the following stub:

```
var Stack:
    external module
        exports(Push,Pop)
        procedure Push(i: SignedInt) = external
        procedure Pop(var i: SignedInt) = external
    end module
```

The stub for a module or monitor can contain declarations of manifest constants, collections, types, and converters; these can be exported and/or used to define the types of the parameters and results of exported functions and procedures.

When linking with a separately compiled module, the module's stub must occur inside exactly one other module (typically the main module) that is being linked. The occurrence of the stub marks the place the separately compiled module would be located, had it not been separately compiled. At that location a call is emitted to execute the module's initialization code.

LINKING PROGRAMS UNDER UNIX

Combining pre-compiled CE programs is easy under Unix. We will illustrate the method using our stack example. Assuming that the stack module (implementation, NOT stub) is in a file named stk.e, we produce its object module by giving the cec command the c flag, as in:

```
cec -c stk.e
```

This produces the object module in the file stk.o. As has been described before, the ".o" suffix means a linkable object file while the ".out" suffix means an executable load module.

Next, we place the stub for the stack module in the file "stack.e". A program that uses the stack module should contain the line:

```
include 'stack.e'
```

to get a copy of stack's stub. Our earlier example, called Reverse, contains exactly this line, and can be compiled as follows:

    cec -c reverse.e

This creates the linkable object module reverse.o, which has external references to the stack module. We can link reverse.o and stk.o to create reverse.out by typing:

    cec reverse.o stk.o

The cec command recognizes the ".o" suffixes, bypasses compilation and proceeds to link the two modules. The name of the resulting executable object module is derived from the first file name in the list.

In general, the cec command compiles files with the ".e" suffix, and links the result with ".o" files, so the line

    cec reverse.e stk.o

compiles reverse.e and links it with stk.o to create reverse.out.

If a program has been written in assembly language or the language C then it is straightforward to link it with CE. Suppose the object module for the program is in file prog.o. All that is required for linkage with a CE program is to include prog.o in the list of files given to the cec command. Details of the conventions for linking CE to/from assembler or C are described in an appendix.

CHAPTER 3 SUMMARY

This chapter has presented the sequential part of Concurrent Euclid (CE), which is known as Sequential Euclid (SE). SE is an important language in its own right, and is used for implementing sequential programs such as the CE compiler. The next chapter introduces the concurrency features of CE. CE is a language designed to support high performance, highly reliable software. The language disallows dangerous or unlikely constructs. At the same time, generated code for CE is quite efficient. The sequential constructs of CE presented in this chapter are now summarized.

Basic data types - These are integers (ShortInt, SignedInt, UnsignedInt, LongInt and AddressType), Booleans, characters and pointers.

Structured types - These are arrays, records and sets.

Strings - A string is a packed array of characters with lower bound one.

Literals - There are integer literals (e.g., 10, 17#8, 0E8#16), character literals (e.g., $H, $$T) and string literals (e.g., 'Hello$N$E').

Constants - In CE, a constant is a value that does not change in a given scope. Assuming x and y are variables, "average" is computed at runtime, when the begin statement is entered, and does not change up to "end":

```
begin
    const average: SignedInt := (x+y) div 2
    ...
end
```

A constant whose value is known at compile time is said to be *manifest.*

I/O package - This is a module that supports input/output such as:

IO.PutString('Hello$N$E') {Prints: Hello}

Parameters - Procedures have parameters that are var or constant (nonvar). Var formal parameters can be assigned to, and the change occurs in the corresponding actual parameter. Constant parameters cannot be changed. The parameters of a function are required to be constant.

Modules - These are syntactic units that control visibility and enforce modularity. From outside a module, only its exported components can be accessed.

Absolute address variables - The "at" clause can be used to place variables at absolute locations. For example,

var ttyData (at 9001#16): Char

places ttyData at hexadecimal location 9001.

Side effects - An expression has a side effect when it changes the value of a variable. Functions and expressions in CE do not have side effects.

Collections - A collection is much like an array. It differs from an array in that its elements are individually allocated (by New) and de-allocated (by Free). Here p is a pointer that is used as a subscript of collection c:

```
var c: collection of LongInt
var p: ^c
c.New(p)
c(p) := 14
```

Return statements - The value of a function must be returned explicitly, as in the statement

```
return(a+b)
```

A return statement without the parenthesized result expression can be used to return from a procedure.

Aliasing - This means having more than one name for a variable. CE prevents aliasing by placing constraints on var parameters and on binding.

Binding - A special declaration is used for re-naming a (part of a) variable. For example, v(i) is renamed x:

```
bind var x to v(i)
```

To avoid aliasing, the name v becomes invisible until the end of x's scope.

Type converters - These have an appearance much like functions. They are used to over-ride the strong type checking of CE.

CHAPTER 3 BIBLIOGRAPHY

The philosophy behind full Euclid [Lampson 1977] is well described by Popek et al [1977,1981]. Welsh et al [1977] describe ambiguities and insecurities of Pascal, which are cleaned up in the design of Euclid. The implementation of the Toronto Euclid subset of Euclid is described by Holt et al [1981]. Guttag [1980] describes how to use Euclid modules to implement abstract data types and how to prove Euclid modules correct. Ottawa Euclid is an extension of Toronto Euclid intended for formal specification and verification [Crowe 1981].

Crowe, David R. Ottawa, Euclid language specification, Report TR-5613-81-7, I.P. Sharp Associates, 156 Front Str. W., Toronto, Canada, November, 1981.

Guttag, J.V. Notes on type abstraction (version 2). IEEE Transactions on Software Engineering, vol. SE-6, no. 1 (Jan. 1980), 13-23.

Holt, R.C., Wortman, D.B., Cordy, J.R., Crowe, D.R., Griggs, I.H., Euclid: A language for producing quality software. Proceedings of National Computer Conference, Chicago, May 1981.

Lampson, B.W., Horning, J.J., London, R.L., Mitchell, J.G., Popek, G.J., Report on the programming language Euclid. SIGPLAN Notices, 12,1 (February 1977). The revised language is described by report CSL-81-12 (Oct. 1981), Xerox Palo Alto Research Center, 3333 Coyote Hill Road, Palo Alto, CA, 94304.

Popek, G.J., Horning, J.J., Lampson, B.W., Mitchell, J.G., London, R.L., Notes on the design of Euclid. Proceedings of ACM Conference on Language Design for Reliable Software, SIGPLAN Notices 12,3 (March 77), 11-18.

Welsh, J., Sneeringen, W.J., Hoare, C.A.R., Ambiguities and insecurities in Pascal. Software Practice and Experience 7,6 (Nov.-Dec. 1977), 685-696.

CHAPTER 3 EXERCISES

1. Write a CE program called Number that reads and prints lines of text, with the number of each line printed on the left. Assume that each line is ended by a new line character ($$N) and that an end-of-file character ($$E) occurs at the end.

2. Write a CE function that is passed a string of digits and returns the positive integer value of the string. Assume the string is terminated by a non-digit character. Hint: the expression Ord(ch)-Ord('0') gives the numeric value corresponding to digit character ch.

3. Consider the following:

```
const c: array 1..10 of SignedInt := (1,2,3,4,5,6,7,8,9,10)
var a: array 1..10 of SignedInt := c
procedure Sum(var m: SignedInt) =
```

```
        imports(a)
        begin
            var i: 1..10 := 1
            m := 0
            loop
                m := m + a(i)
                exit when i = 10
                i := i + 1
            end loop
        end Sum
    ...
Sum(a(5))
```

What is the apparent purpose of Sum? If this program is translated to an equivalent one in Pascal, what would be the value of a(10) after the call to Sum. What restriction of CE makes this program illegal? Why does CE have this restriction?

4. Consider the following:

```
const s: array 1..5 of SignedInt := (1,2,3,4,5)
function GetNext(var i: SignedInt) returns r: SignedInt =
        imports(s)
        begin
            i := i + 1
            return(s(i))
        end GetNext
function GetPrevious(var i: SignedInt) returns r: SignedInt =
        imports(s)
        begin
            i := i - 1
            return(s(i))
        end GetPrevious
var i: SignedInt := 2
var a: array 1..10 of SignedInt
a(i) := GetNext(i) + GetPrevious(i)
```

What is the apparent purpose of the GetNext function? In most languages (Pascal, Fortran, C, PL/I, etc) the order of evaluation of functions in an expression is not specified. Given a translation of our program into one of those languages, what are the possible different effects of the assignment statement? What restriction of CE makes the program illegal? Why does CE have this restriction?

5. Write a CE module that has entries Allocate and DeAllocate to manage the blocks on a disk. Allocate is called to get a presently unallocated block and DeAllocate is called to give back an allocated block. Assume the blocks are numbered 0 to maxDiskBlock. Use a bit map to manage the blocks. This bit map can be implemented as an array of sets, each set ranging from say, 0 to 15. A bit will be on when the block has been allocated. Always allocate the available block with the smallest block number. If no more blocks are available, Allocate returns the value nullBlock, which is maxDiskBlock + 1.

6. Write a disk block manager module as per the previous exercise, but provide a new entry called HeadPosition. The procedure HeadPosition does nothing but inform the module where the read-write head of the disk is positioned. Using this information, minimize head motion by allocating the

block nearest the present head position.

7. Write a CE procedure that produces a dump of a certain area of memory. Your procedure has the stub:

```
const binary := 2
const octal := 8
const decimal := 10
const hexadecimal := 16
procedure Dump(location: AddressType,
    areaSize: AddressType,
    numberBase: ShortInt) = external
```

The procedure prints out the value of areaSize bytes using the specified base, such as octal. Hint: The Peek function given in this chapter accesses arbitrary bytes in memory.

8. Implement a module called Logger, described as follows:

```
pervasive const trace := 0
pervasive const profile := 1
pervasive const silence := 2
var Logger:
    external module
        exports(Log, Report, Option)
        procedure Log(action: UnsignedInt,
            message: packed array 1..parameter of Char) =
            external
        procedure Report = external
        procedure Option(setting: trace..silence) =
            external
```

end module

The Log procedure is called to log (record) an event, which is described by an event number and a string. The Option procedure sets the option to be trace, profile or silence. When the option setting is trace, all actions are printed as logged. When Report is called, if the option is silence, then the most recent ten actions are printed, and if the profile option is set, a count of the occurrences of each action number is printed.

Chapter 4

CONCURRENT EUCLID: CONCURRENCY FEATURES

This chapter introduces the concurrency features of Concurrent Euclid. These features are processes (to support concurrent activity), monitors (to gain mutually exclusive access to data) and signal/wait statements (to allow processes to block and to be waked up). There is also the busy statement, which allows us to use CE as a simulation language.

SPECIFYING CONCURRENCY

We begin with a simple example of two processes, named Hi and Ho, that repeatedly print their own names. As can be seen, a process in CE is written as the keyword "process" followed by the name of the process (Hi or Ho in this example); then the rest of the process is like a procedure.

```
var HiHo:
    module
        include '%IO1'

        process Hi
            imports(var IO)
            begin
                loop
                    IO.PutString('Hi$N$E')
                end loop
            end Hi
```

```
process Ho
    imports(var IO)
    begin
        loop
            IO.PutString('Ho$N$E')
        end loop
    end Ho
```

end module

The Hi and Ho processes execute in parallel at undefined relative speeds. The output of the program is an unpredictable sequence of Hi's and Ho's:

```
Ho
Ho
Hi
Ho
Hi
Hi
Hi
Hi
Ho
...etc...
```

The order of Hi's and Ho's may vary from execution to execution. It is even possible that a part of the string Hi (just the H) might be printed, then Ho and then the rest of Hi. We have not shown this possibility; we have assumed that the IO module has been implemented such that it completely prints one string before starting another.

In CE, any module can contain processes. These must appear at the end of the module, following the module's "initially" procedure (if present). After the module has been initialized, all these processes begin to execute. Note that processes in CE are similar to the cobegin/end construct described in Chapter 2.

RE-ENTRANT PROCEDURES

Sometimes several processes execute similar algorithms. For example, in a computer configuration with three terminals, the operating system may use three almost identical processes to manage these terminals. In CE, processes can share algorithms by calling the same procedure.

We can factor out the common parts of Hi and Ho from our previous example, creating a procedure called Speak:

```
var HiHo2:
    module
        include '%IO1'

        procedure Speak(word: packed array 1..4 of Char) =
            imports(var IO)
            begin
                loop
                    IO.PutString(word)
                end loop
            end Speak

        process Hi
            imports(Speak)
            begin
                Speak('Hi$N$E')
            end Hi

        process Ho
            imports(Speak)
            begin
                Speak('Ho$N$E')
            end Ho

    end module
```

HiHo2 behaves like the old HiHo module, but now there is only one copy of the loop and the call to PutString. The Speak procedure is *re-entrant*, meaning that it can be called concurrently by several processes. In CE, all procedures and functions are re-entrant. Parameters, such as the string called "word" and local variables, are distinct from process to process. For example, Hi's activation of Speak has word set to 'HiNE' while Ho's activation of Speak has word set to 'HoNE'.

Re-entrant procedures are essential for software such as operating systems because (1) they make the software easier to understand, and (2) they make the code smaller.

MUTUAL EXCLUSION

When processes need to update common data, the data may be corrupted if more than one update takes place in parallel. In the preceding chapters, *monitors* were introduced as a feature for guaranteeing mutually exclusive access to common data. A monitor can be considered to be a fence around the data; all code accessing the data is gathered into procedures/functions and moved inside the fence. Processes wishing to access the data do so by entering a *gate* or *entry* in the fence, to execute one of these procedures or functions. The monitor guarantees that only one process is active inside the fence at a given time.

Chapter 2 illustrates the danger of concurrently updating data by an example involving observer and reporter processes. The following program shows how this example can be safely programmed using a monitor.

The CE program contains two processes, named Observer and Reporter, and a monitor named Update. The shared data is the variable called count. The monitor is initialized before the processes are "born", so count starts out with the value zero.

```
var Counting:
    module
        include '%IO1'

        var Update:
            monitor
                imports(var IO)
                exports(Observe, Report)

                var count: SignedInt := 0

                procedure Observe =
                    imports(var count)
                    begin
                        count := count + 1
                    end Observe

                procedure Report =
                    imports(var IO, var count)
                    begin
                        IO.PutInt(count, 8)
                        count := 0
                    end Report
```

```
        end monitor

    process Observer
        imports(var Update)
        begin
            ...
            Update.Observe
            ...
        end Observer

    process Reporter
        imports(var Update)
        begin
            ...
            Update.Report
            ...
        end Reporter

    end module
```

The Observer process calls the entry Update.Observe, which adds one to count. The Reporter process calls Update.Report, which prints the value of count and sets it to zero. If the Reporter tries to enter the monitor while the Observer is inside the monitor, the Reporter is held up at the entry point until the Observer leaves the monitor. Similarly, the Observer is prevented from entering the monitor while the Reporter is inside. The mutexbegin/end construct, described in Chapter 2, is an analogous feature for guaranteeing mutually exclusive access to shared data.

The data in a monitor (count in this example) is static in that it remains intact from invocation to invocation of the monitor. By contrast, the entries to the monitor are re-entrant procedures/functions, and their parameters and local variables are private to the invoking process. These parameters and local variables are allocated when the procedure/function is called and de-allocated upon return.

A monitor in CE is similar in form to a module and can have an "initially" procedure. This initially procedure is executed before any processes enter the monitor. A monitor cannot contain modules, monitors or processes. Scope rules for monitors are the same as those for modules: the monitor and its procedures/functions must import any identifiers they use. The only internal parts of the monitor that are externally visible are those names that the monitor exports. A monitor is not allowed to export variables. An entry to a monitor must not be called from inside a monitor.

WAITING AND SIGNALING

It is common that processes must synchronize their activities. A typical situation is that processes compete for shared resources. Once a resource is allocated to one process, another process needing the resource should be blocked until the first process releases it. The following monitor allocates a single resource among processes sharing it. Each process "acquires" the resource, then uses it, and finally "releases" it. In the Acquire procedure, the process blocks itself by the *wait* statement if the resource is not available. In the Release procedure, the *signal* statement wakes up one waiting process (if there is one). If there is no waiting process, the signaler just continues.

```
var Resource:
    monitor
        exports(Acquire, Release)

        var inUse: Boolean := false
        var available: condition {Signaled when not inUse}

        procedure Acquire =
            imports(var inUse, var available)
            begin
                if inUse then
                    wait(available)
                end if
                inUse := true
            end Acquire

        procedure Release =
            imports(var inUse, var available)
            begin
                inUse := false
                signal(available)
            end Release

    end monitor
```

The signal and wait statement apply to *conditions*, which are queues of processes. A process that executes a wait statement is blocked and "steps out" of the monitor until a signaling process wakes it up. When a process executes a signal statement, the corresponding condition queue is checked. If it contains processes, one is removed and allowed to continue immediately. The signaler "steps out" of the monitor and is not allowed to continue until no more processes are in the monitor. If the signaled condition has no processes waiting for it, no process is awakened, and the signaler continues to execute. However, before the signaler continues, other processes may enter and leave the monitor. Note that at most one process is active in the monitor, so mutually exclusive access to the monitor's data is guaranteed.

The signal and wait statements of CE are analogous to the block and wakeup statements described in Chapter 2. The Acquire and Release entries of the Resource monitor behave like P and V operations on a binary semaphore. This illustrates the fact that monitors are as powerful as semaphores, and that they can be used to implement semaphores.

DETAILS OF SIGNALING, WAITING AND CONDITIONS

We will now explain signal/wait statements and conditions in more detail. A condition is a queue of processes, which is implicitly initialized to be empty. A wait statement adds a process to the queue and stops the process from executing. A signal statement removes a waiting process and causes it to continue execution immediately. The signaling process is put to sleep until no other process is active in the monitor.

The reason the awakened process proceeds before the signaling process is so that the awakened process can be sure that the situation it was waiting for is true. It is the responsibility of the signaler to guarantee that the desired situation exists before doing the signal.

We can test to see if a condition queue is empty, as in this example.

if empty(available) then ...

Empty is a pre-defined function used to determine if there are processes waiting for a condition.

The signal statement wakes up one process (when one or more processes are waiting on the condition). If no processes are waiting on the condition, the signaler simply continues. However, before continuing, the signaler may step out of the monitor and allow other processes to enter it.

The only place a condition can be declared is as a field of a monitor. We can have arrays of conditions. Conditions cannot be assigned, compared or passed as parameters.

The scheduling for conditions is *fair*, meaning that given enough signals, every waiting process will eventually be waked up. It is typical for an implementation to use FIFO scheduling, which is obviously fair. However, in general we are not guaranteed that the order is FIFO.

ASSERT STATEMENTS

Along with condition declarations, we generally give a comment specifying the awaited situation. For example, the declaration of the inUse condition specifies that it is signaled when "not inUse". The situation is expressed as a logical relation among the values of monitors variables. In the Concurrent Pascal language a condition is called a "queue", but CE uses the term "condition" to emphasize that a particular situation (condition) is being waited for.

A common usage pattern for conditions is to check to see if a situation exists and if not, to wait for it, as in:

```
if inUse then
    wait(available)
end if
assert(not inUse)
```

Here we have added an assert statement to document the fact that once we pass the "end if" we know that the resource is "not inUse".

The assert statement can be thought of as an "executable comment". It specifies a relationship that is necessary for the program to be correct. By default, the relationship is checked at run time and the program is aborted if it is not true. The overhead required by this check can be eliminated by a compiler option that effectively tells the compiler to ignore assert statements. Typically a program is tested with the checking active; later, if performance is critical, the program is re-compiled to remove the run-time checking.

PRIORITY CONDITIONS

There is another kind of condition that allows processes to be waked up in a specified order. We call these *priority* conditions. They are used just like ordinary conditions except that a wait on a priority condition must specify a non-negative SignedInt priority, for example:

```
var c: priority condition
...
wait(c, 13)
...
signal(c)
```

A signal of priority condition c will wake up the waiting process (if any) with the lowest specified priority. Since priorities must be non-negative, processes with zero priority will be waked up first.

Scheduling for priority conditions is *not* fair in that a process with a large valued priority may be indefinitely overtaken by a sequence of waits by processes with low numbered priorities.

AN EXAMPLE PROGRAM: MANAGING A CIRCULAR BUFFER

We will use monitors with signaling and waiting to solve a typical problem in concurrent software. Two processes, a producer and a consumer, are communicating by means of a shared buffer. The producer sends data (messages) to the consumer. The buffer is used as a queue to hold these messages, so either process may occasionally slow down without impacting the speed of the other. This queue will be managed in FIFO order so that messages are always received in the same order they were sent.

This problem can be efficiently solved using a multi-slot buffer, where each slot can hold one message. The slots are re-used cyclically and hence the arrangement is called a *circular buffer*. The producer fills slots, stopping when there are no more free slots. The consumer empties slots, stopping when there are no more full slots.

We will use a monitor, called MailBox, with entries Send and Receive. The following variables will represent the queue:

buffer:	array 1..numberSlots of MessageType
numberFull:	0..numberSlots {How many full slots}
slotToFill:	1..numberSlots {Slot for next send}
slotToEmpty:	1..numberSlots {Slot for next receive}

Waiting occurs in the monitor for two reasons: (1) when all slots are full, the producer must wait for a slot to become empty, and (2) when all slots are empty, the consumer must wait for a slot to become full. These awaited situations are represented by the two conditions:

var emptySlot:	condition {When numberFull < numberSlots}
var fullSlot:	condition {When numberFull > 0}

The Send entry of MailBox tests to see if there is a free slot; if not, it waits for one. Next, it adds the new message to the queue. Finally, it signals the fullSlot condition, in case the consumer is waiting for a message. The Receive entry is analogous: it conditionally waits for a full slot, empties the slot and signals the emptySlot condition.

```
var MailBox:
    monitor
        imports(MessageType)
        exports(Send, Receive)
```

```
pervasive const numberSlots := 5
var buffer: array 1..numberSlots of MessageType
var numberFull: 0..numberSlots := 0 {Full slots}
var slotToFill: 1..numberSlots := 1 {Slot for send}
var slotToEmpty: 1..numberSlots := 1 {Slot for receive}
var emptySlot: condition {When numberFull<numberSlots}
var fullSlot: condition  {When numberFull>0}

procedure Send(msg: MessageType) =
    imports(var buffer, var numberFull, var slotToFill,
        var emptySlot, var fullSlot)
    begin
        if numberFull = numberSlots then
            wait(emptySlot)
        end if
        assert(numberFull < numberSlots)
        buffer(slotToFill) := msg
        slotToFill := (slotToFill mod numberSlots) + 1
        numberFull := numberFull + 1
        signal(fullSlot)
    end Send

procedure Receive(var msg: MessageType) =
    imports(buffer, var numberFull, var slotToEmpty,
        var emptySlot, var fullSlot)
    begin
        if numberFull = 0 then
            wait(fullSlot)
        end if
        assert(numberFull > 0)
        msg := buffer(slotToEmpty)
        slotToEmpty := (slotToEmpty mod numberSlots) + 1
        numberFull := numberFull - 1
        signal(emptySlot)
    end Receive

end {MailBox} monitor
```

The MailBox monitor buffers messages of whatever type is specified by MessageType. Assuming that this monitor is in the file mailbox.e, we can test it with MessageType set to Char as follows:

```
var TestMailBox:
    module
        type MessageType = Char
        include 'mailbox.e'
        include '%IO1'

        process Producer
            imports(var IO, var MailBox)
            begin
                var c: Char
                loop
                    IO.GetChar(c)
                    MailBox.Send(c)
                    exit when c = $$E
                end loop
            end Producer

        process Consumer
            imports(var IO, var MailBox)
            begin
                var c: Char
                loop
                    MailBox.Receive(c)
                    IO.PutChar(c)
                    exit when c = $$E
                end loop
            end Consumer

    end {TestMailBox} module
```

Our test uses messages that are single characters. It copies the input stream into the output stream until an end-of-file character is found.

This example illustrates the fact that monitors can be used to implement message passing (described in Chapter 2).

SIMULATION MODE AND KERNELS

A CE program, such as our test of the MailBox monitor, can be run as an ordinary job under an operating system. This type of execution is called *simulation mode*. It is supported by a *simulation kernel* that shares the job's CPU time among the processes of the CE program. Simulation mode is particularly valuable for initial testing of low level software, such as device drivers. Eventually this software will be run without operating system support using a *bare machine kernel* that handles interrupts and shares CPU time among processes. The next section shows how a device driver can manage peripheral I/O devices using a bare machine kernel. The last chapter of the book explains how kernels are implemented.

BASIC DEVICE MANAGEMENT

In the last chapter, we gave an example of driving a device, namely the keyboard of a terminal. This was done by having the processor do a busy wait until the input/output was completed. This wastes processor power and is unacceptable in many situations.

To avoid busy waiting, one can use procedures called BeginIO, WaitIO and EndIO. These procedures are not part of the CE language, but are supported by the bare machine kernel. This will be explained by an example from Tunis, a Unix-compatible operating system written in CE. Its driver for reading from terminal number 1 has this form:

```
loop
    TtyDoIO.GetChar1 (c)
    TerminalMonitor.BufferChar1 (c)
end loop
```

This repeatedly gets character c from the device and calls a monitor to store the character in a buffer.

The procedure GetChar1 is machine dependent and so it is isolated in a module called TtyDoIO; it can be written as follows for a PDP-11.

```
procedure GetChar1 (var c: Char) =
    imports(BeginIO, WaitIO, EndIO)
    begin
        const inputCharReady := 7
        var tty1InputStatus (at 176500#8): set of 0..7
        var tty1InputBuffer (at 176502#8): Char
        BeginIO
            if inputCharReady not in tty1InputStatus then
                WaitIO(tty1InputId)
            end if
            c := tty1InputBuffer
        EndIO
    end GetChar1
```

This procedure uses device registers at octal locations 176500 and 176502; the first gives the status of the device and the second holds the most recently received character from the keyboard. The BeginIO entry of the kernel disables all interrupts. The WaitIO procedure blocks the executing process until the read operation is complete, i.e., until the device interrupts the CPU. During the wait, the interrupts are re-enabled and other processes are executed. When WaitIO returns, the next character from the keyboard is in tty1InputBuffer; the interrupts are again disabled and remain disabled until EndIO is called. The execution of the WaitIO is conditional in this procedure to avoid the overhead of waiting when the next character is already available.

This approach to basic device management, using Begin/Wait/EndIO, provides a fast, clean interface to devices. Of course, basic procedures such as GetChar1 are machine dependent and must be replaced when porting to a new machine. Fortunately they can be written in CE for machines like the PDP-11 with memory mapped device registers, such as tty1InputStatus. But for non-mapped architectures such as the IBM System/360, support of new devices entails modifying the kernel to emit explicit "Start IO" machine instructions.

SIMULATION AND THE BUSY STATEMENT

CE provides a statement called "busy" that allows us to simulate the passage of time when running in simulation mode. Suppose we are testing low level software such as the device managers of the Tunis operating system. We can execute these programs in simulation mode by replacing those parts that activate the I/O devices. For example, we would replace the TtyDoIO module by an equivalent module or monitor that uses files to simulate terminals. Each I/O device can be simulated by a process; for example, the following simulates the printer of terminal number one:

```
process TtyPrint1 {Simulate printer of a teletype}
    imports(var TtyDoIO, var IO)
    begin
        var c: Char
        loop
            TtyDoIO.FetchToPrint1(c)
            busy(10) {Simulate 10 time units to print char}
            IO.PutChar(c)
        end loop
    end TtyPrint1
```

The busy statement causes this process to be held up until the specified number of units of simulated time have passed. The size of a unit of time is arbitrary and is chosen by the programmer. In this simulation our unit is a hundredth of a second, and we are simulating an old-style teletype that takes about a tenth of a second to print a character. In the next section we will discuss simulated time in more detail.

The TtyDoIO module will be replaced by a monitor that has entries named PutChar1, GetChar1, etc., that simulate output and input to the terminals. PutChar1 places its character in a buffer named outBuf1, and FetchToPrint1 removes the character.

```
var TtyDoIO: {Simulates interface to tty hardware}
    monitor
        exports(GetChar1, PutChar1, ..., FetchToPrint1, ...)
        ...
        var outBuf1: Char
        var outBufFull1: Boolean := false
        var fullOutBuf1: condition {when outBufFull1}
        var emptyOutBuf1: condition {when not outBufFull1}

        procedure PutChar1(c: Char) =
            imports(var outBuf1, var outBufFull1,
                var fullOutBuf1, var emptyOutBuf1)
            begin
                if outBufFull1 then
                    wait(emptyOutBuf1)
                end if
                assert(not outBufFull1)
                outBuf1 := c
                outBufFull1 := true
                signal(fullOutBuf1)
            end PutChar1

        procedure FetchToPrint1(var c: Char) =
            imports(outBuf1, var outBufFull1,
                var fullOutBuf1, var emptyOutBuf1)
            begin
                if not outBufFull1 then
                    wait(fullOutBuf1)
                end if
                assert(outBufFull1)
                c := outBuf1
                outBufFull1 := false
                signal(emptyOutBuf1)
            end FetchToPrint1
```

...

end module

As you may have noticed, this monitor implements a simple mailbox where the producer calls PutChar1 and the consumer (TtyPrint1) calls FetchTo-Print1. The algorithm is very similar to the one used in our circular buffer implementation of mailboxes.

Since there is only one slot in the buffer, the producer and consumer do not execute in parallel. They execute in strict alternation, like alternate runners in a relay race. The monitor solves the *baton passing problem*, which is the problem of passing control from one process to another and back repeatedly. From a conceptual point of view, the producer and consumer act as a single process (a coroutine) that produces, then consumes, then produces, and so on.

If there are several terminals to be simulated, we would make PutChar1 into PutCharN and pass an extra parameter giving the terminal number. FetchToPrint1 would similarly be modified. OutBuf1 and OutBuf-Full1 would become arrays indexed by terminal numbers. Similarly the two conditions would become arrays. In general we can use arrays inside a monitor to make the monitor behave like an array of monitors.

SIMULATED TIME AND PROCESS UTILIZATION

The busy statement causes a process to be blocked for a given amount of simulated time. Simulated time can be understood in terms of a special clock maintained by the simulation kernel. Before a CE program begins executing, this clock is set to zero. It operates (ticks) only as a result of processes executing busy statements. As long as processes are executing other statements, the clock does not change. What this means is that the (simulated) time to execute statements other than busy is considered to be negligible. Of course the programmer can cause his code to use simulated time by sprinkling in busy statements at appropriate places.

The simulation kernel can implement the busy statement and update the simulated clock in the following manner. A queue of busy processes is kept in ascending order of the simulated time at which the process is to wake up from executing "busy". If at simulated clock time t, a process executes a busy statement with parameter b, then its time to wakeup is $t+b$; this value is used for merging the process into the queue.

When no more processes are ready to execute (due to blocking by busy, by wait or by trying to enter a monitor), the first process on the busy queue is made ready and the clock is set forward to the process's wakeup time. This is the only circumstance in which the clock is advanced. As this process executes, it may in turn wake up other processes (via signaling or exiting monitors) thereby making them ready, and it will probably eventually become blocked. The manipulation of the busy queue and the clock continues in this manner to simulate the passage of time.

The busy statement implies an ordering among the execution of processes, and it allows us to compute the *utilization* of each process. This utilization is defined as the fraction of the total simulated time that the process was held up at a busy statement. The idea is that useful work done by the process is simulated by execution of busy statements. To understand this, look at the process named TtyPrint1 that simulates the printer of a teletype. The simulated printer consumes ten units of time whenever it prints a character. If the simulated printer runs at full speed then the only thing that slows it down (in simulated time) is the execution of busy, and so the process's utilization is 100%. But if the process is held up in the TtyDoIO monitor waiting for another character, and simulated time passes, then its utilization drops. At a minimum the utilization of a process is zero per cent, meaning it has executed no busy statements.

PROCESS STATISTICS

The simulation kernel for CE under Unix produces process statistics, including the utilization of each process. For example, when a CE program is stopped (by a quit character) under Unix a table similar to the following is printed:

process	maxmem	%util	state	
1	370	29	ready	at line 27 of file 1
2	222	84	blocked	at line 49 of file 1

The numbers in the left column give the number of the process, counting processes in textual order. The percent utilization is as described in the previous section. The state of the process is given: at the time the example program was stopped, process 1 was ready and process 2 was blocked. The statement being executed by each process is given in terms of the line number within source file, where files are counted starting with one for the main module and increasing by one for each included file.

The column labeled maxmem tells how much space was used for the process's local data. This space is allocated on the process's stack and includes space for parameters and local variables that are allocated at procedure/function entry and de-allocated upon return. There is a default amount of space, 2000 bytes in a typical implementation, set aside for each process's stack. If this is too small, or overly generous, then the amount can be explicitly specified. For example, only 400 bytes are specified here:

process TtyPrint1 (400)

The actual amount used by the process is recorded under "maxmem" and can be used as a guide for explicitly specifying the size.

This concludes our introduction of the concurrency features of CE. We have given basic examples that handle problems of mutual exclusion and process synchronization. The next chapter gives more examples of CE programs that solve concurrency problems.

CHAPTER 4 SUMMARY

In this chapter we introduced the concurrency features of CE. Here is a summary of these features.

Processes - These execute in parallel at undefined relative speeds. Processes can call procedures and functions, which are re-entrant in CE. The process header can specify the size of stack needed for the process's local data.

Monitors - These are fences enclosing critical data shared among processes. This data inside a monitor is static and retains its values between calls to the monitor. A monitor has the same form as a module, but cannot contain internal modules, monitors or processes. Modules and monitors are initialized in textual order in the program.

The wait statement - A wait statement is executed in a monitor to block the process and remove it from the monitor.

The signal statement - A signal statement is executed in a monitor to specify that an awaited situation exists. The signal immediately wakes up one waiting process (if any are waiting). The signaling process is blocked until no other processes are active in the monitor.

Conditions - These are the operands of wait and signal statements. Conditions represent awaited situations. Priority conditions require a non-negative priority parameter for wait; signaling a priority condition wakes up low priority processes first.

Empty - This is a pre-defined function that tests if there are processes waiting for a condition.

The busy statement - This statement is used in simulation mode to simulate the passage of time. A process executing busy (b) is delayed until b units of simulated time have passed.

Process statistics - The simulation kernel gathers statistics during program execution and prints them at program termination. These statistics include process utilization (percentage time in busy statements) and stack usage.

The example CE programs in this chapter solve several interesting problems. The Resource monitor shows how a single resource can be managed and how monitors can implement semaphores. The MailBox monitor shows how a circular buffer is used to implement message passing and Send and Receive operations. The Begin/Wait/EndIO operations supported by a bare machine kernel provide a clean, efficient method of driving peripheral devices. The purpose of a DoIO module, such as TtyDoIO, is to isolate low level machine dependencies, so the rest of the software is relatively portable and can be tested in simulation mode. The simulation version of TtyDoIO solves the baton passing problem, in which one process runs then blocks and passes control to another process, and so on as in a relay race.

CHAPTER 4 BIBLIOGRAPHY

The monitor as a language feature was invented when it was realized that concurrent programs are simplified by gathering code accessing critical data into one place. See Hoare and Perrott's book [1972, pages 91 and 109] for early discussions of monitor-like constructs.

Concurrent Pascal was the first language to incorporate monitors, and Brinch Hansen's articles and books listed below are important references. The "queues" of Concurrent Pascal are like the conditions of CE except that at most one process at a time is allowed to be waiting for a queue, and a signaling process implicitly returns from the enclosing monitor. Concurrent Pascal is notable in that its compiler prohibits processes from accessing shared data except in monitors. This is not the case in CE. Brinch Hansen's new language, Edison[1981], does not prevent processes from sharing data.

Hoare [1974] formalized monitors, giving proof rules for signal and wait. He also gave examples of monitors and the implementation of monitors in terms of semaphores. It appears that SUE/11 was the first production operating system to be based on monitors [Greenblatt 1976].

Wirth's Modula language [1977] has monitor-like constructs. His later language Modula-2 [1980] does not incorporate monitors as a language feature; it provides more primitive constructs from which monitors can be implemented. The book by Welsh and McKeag [1980] gives examples of the use of monitors in the Pascal Plus language.

The book you are reading evolved from the book *Structured Concurrent Programming with Operating Systems Applications* [Holt et al 1978]. That book uses Concurrent SP/k (CSP/k), which is a PL/I subset (SP/k) extended with processes and monitors.

Brinch Hansen, P. Structured multiprogramming. *Comm. ACM 15*, 7 (July 1972), 574-577.

Brinch Hansen, P. The programming language Concurrent Pascal. *IEEE Trans. on Software Engineering SE-1*,2 (June 1975), 199-207.

Brinch Hansen, P. *The Architecture of Concurrent Programs.* Prentice-Hall (1977).

Brinch Hansen, P. Edison: a multiprocessor language. *Software Practice and Experience 11*,4 (April 1981), 325-361.

Greenblatt, I.E. and Holt, R.C. The SUE/11 operating system. *INFOR*, Canadian Journal of Operational Research and Information Processing *14*,3 (October 1976), 227-232.

Hoare, C.A.R. Monitors: an operating system structuring concept. *Comm. ACM 17*, 10 (October 1974), 549-557.

Hoare, C.A.R. and Perrott, R.M. (editors) *Operating Systems Techniques*, Academic Press, 1972.

Holt, R.C., Graham, G.S., Lazowska, E.D., Scott, M.A., *Structured concurrent programming with operating systems applications.* Addison-Wesley, 1978.

Welsh, J. and McKeag, M. *Structured System Programming.* Prentice-Hall 1980.

Wirth, N. Modula: a language for modular programming. *Software Practice and Experience Vol. 7*, 1 (January-February 1977), 3-35.

Wirth, N. Modula-2. Report number 36, Institut fur Informatik, Eidgenossische Technische Hochschule, Zurich, March 1980.

CHAPTER 4 EXERCISES

1. Re-write the Speak procedure of the HiHo2 program so that the characters of each word are not necessarily printed contiguously. Hint: break up PutString into components.

2. Suppose there were several reporter processes using the Update module in the counting program. Will any events be lost or multiply reported? Explain why or why not.

3. Write a program that generates sentences of the form < subject,verb,object > as follows. There is a phrase allocator monitor with entries Acquire and Release. These entries have two parameters: phrase type (subject, verb or object) and a string (the phrase). There are several sentence generator processes that repeatedly acquire a subject, a verb and an object, print the resulting sentence and release the three phrases. A sentence printing monitor has a single entry; it accepts three strings (subject, verb and object) and prints the corresponding sentence on a line. The store of phrases managed by the allocator monitor might be as follows:

Subjects	*Verbs*	*Objects*
John Wayne	tamed	the dappled stallion
The green Edsel	generates	a lot of smoke
Ronald Reagan	underfinanced	California universities

4. Write a CE program with two processes that use busy statements. Arrange things so one process has a utilization of 25% and the other has a utilization of 75%.

5. The Resource monitor can be modified to have Acquire accept an extra parameter giving the importance of the requesting process. How would the monitor be changed to give a released resource to the most important process. What is the disadvantage of this arrangement (when the resource is being heavily utilized).

6. Suppose the pre-defined function empty were not present. How would you simulate it using other CE features.

7. For a condition C, implement a parameterless function CPriority that gives the priority of the first process waiting for C. If no processes are waiting for C, the named constant infinity is returned. All waits on C should be via a procedure CWait and all signals via CSignal, which you are to implement. Your implementation should work with a fixed amount of storage

(no arrays) for any number of processes.

8. Implement and run the single resource monitor named Resource in CE with three processes (using the busy statement) and include output statements to show the progress of the processes and the state of the resource.

9. In the Acquire entry of the Resource monitor, why does a process not retest to see if inUse has become false when the process resumes execution following its wait?

10. FIFO was described as a fair scheduling policy for resuming a process waiting on a non-priority condition, in that it did not allow a waiting process to be postponed indefinitely. Are the following scheduling policies fair?

    a) LIFO   (last-in first-out)
    b) LRR   (least-recently-run)
    c) MRR   (most-recently-run)
    d) LFR   (least-frequently-run)
    e) MFR   (most-frequently-run)

11. The monitor concept is at least as powerful as the semaphore concept because it can be used to implement semaphores; this is what the single resource monitor showed. Show the converse: the semaphore concept is at least as powerful as the monitor concept because a monitor can be simulated using semaphores and their associated operations. You should guarantee that mutual exclusion is present in the execution of monitor entries. The equivalent of the signal operation should allow some waiting process (if there is one) to execute in the monitor. Do not support priority conditions.

12. Suppose the Resource monitor is required to enforce an upper bound of N processes accessing the resource, instead of just 1. Exclusive access is not required now, but an upper limit on the number of active processes accessing the resource must be imposed. How should the monitor be changed?

13. What would happen if the Consumer or the Producer process of the MailBox program was omitted?

Chapter 5

EXAMPLES OF CONCURRENT PROGRAMS

This chapter presents solutions in CE to concurrent programming problems of a more substantial nature than those encountered earlier in this book. The purpose is to teach more about both concurrency and CE, through examples. The concurrency issues discussed include synchronization, mutual exclusion, deadlock, and indefinite postponement.

DINING PHILOSOPHERS

Suppose several processes are continually acquiring, using, and releasing a set of shared resources. We want to be sure that a process cannot be deadlocked (blocked so that it can never be signaled) or indefinitely postponed (continually denied a request).

A colorful version of this problem can be stated in terms of a group of philosophers eating spaghetti. It goes like this:

> There are N philosophers who spend their lives either eating or thinking. Each philosopher has his own place at a circular table, in the center of which is a large bowl of spaghetti. To eat spaghetti requires two forks, but only N forks are provided, one between each pair of philosophers. The only forks a philosopher can pick up are those on his immediate right and left. Each philosopher is identical in structure, alternately eating then thinking. The problem is to simulate the behavior of the philosophers while avoiding deadlock (the request by a philosopher for a fork can never be granted) and indefinite postponement (the request by a philosopher for a fork is continually denied).

We will concentrate on the case of five philosophers. Here is a picture of a table setting with five plates and forks.

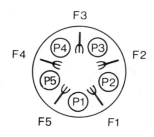

Several points should be clear from the problem description. Adjacent philosophers can never be eating at the same time. Also, with five forks and the need for two forks to eat, at most two philosophers can be eating at any one time. Any solution we develop should allow maximum parallelism.

Consider the following proposed solution. A philosopher acquires his forks one at a time, left then right, by calling a monitor entry PickUp, giving as a parameter the appropriate fork number. Similarly, a philosopher returns his forks one at a time, left then right, by calling a monitor entry PutDown. The philosopher's activity is represented by a process that repeatedly executes the statements:

>   PickUp(left)
>   PickUp(right)
>   Busy eating
>   PutDown(left)
>   PutDown(right)
>   Busy thinking

The entries are part of a monitor that controls access to the forks. The data of the monitor includes the one-dimensional Boolean array idleForks, where idleForks(i) gives the availability of fork number i. Only when a philosopher acquires his two forks does he begin eating. Periods of eating and thinking by a philosopher can be represented by busy statements of appropriate duration.

Unfortunately, this simple solution suffers from a serious defect, namely deadlock. Consider a sequence of process executions in which the philosophers each acquire a left fork, then each attempt to pick up a right fork. Each philosopher will be blocked in the PickUp entry on a condition that can never be signaled; the request for a right fork can never be granted. Deadlock occurs in this situation because processes hold certain resources while requesting others. Clearly, we need a solution that prevents deadlock.

We now discuss a solution that prevents deadlock. Each philosopher is represented by a process that repeatedly executes the statements

>   PickUp(i)
>   Busy eating
>   PutDown(i)
>   Busy thinking

where i is the number of the philosopher. Picking up the forks is now represented as a single monitor entry call.

We will develop a monitor named Forks with two entries, PickUp and PutDown, which acquire and release the forks. The structure of these entries differs from that of the previous entries. The monitor must have variables that keep track of the availability of the five forks. This could be done by having an array of five elements; this is the earlier idleForks method. We will use a different approach. There is still an array of five elements, but the elements will correspond to the five philosophers. The array will be called freeForks, where freeForks(i) is the number of forks available to philosopher i: either 0, 1, or 2. In the PickUp entry, philosopher i is allowed to pick up his forks only when freeForks(i)=2. Otherwise, he waits on the condition ready(i); each philosopher thus has his own condition for which he waits. When he succeeds in picking up his forks, he must decrease the fork counts of his neighbor philosophers. He then leaves the monitor and commences eating.

The neighbors of philosopher i are numbered Left(i) and Right(i). Based on our earlier diagram, Left(3) is 2 and Right(3) is 4. It is important to note that Right(5) is 1 and Left(1) is 5. Left and Right could be implemented as vectors initialized to the appropriate values or as functions using the mod pre-defined function to calculate the proper neighbor number.

```
var Spaghetti: {Solution to the dining philosopher's problem}
    module

        var Forks:
            monitor
                exports(Pickup, PutDown)
                pervasive const numberPhilos := 5
                var freeForks: array 1..numberPhilos of 0..2
                var ready: array 1..numberPhilos of condition
                    {ready(i) when freeForks(i) = 2}

                ...define Left and Right...

                procedure PickUp(me: 1..numberPhilos) =
                    imports(var freeForks, var ready, left, right)
                    begin
                        if freeForks(me) not= 2 then
                            wait(ready(me))
                        end if
                        assert(freeForks(me) = 2)
                        freeForks(Right(me)):=freeForks(Right(me))-1
                        freeForks(Left(me)):=freeForks(Left(me))-1
                    end PickUp
```

```
procedure PutDown (me: 1..numberPhilos) =
    imports (var freeForks, var ready, left, right)
    begin
        freeForks (Right (me)) :=freeForks (Right (me)) +1
        freeForks (Left (me)) :=freeForks (Left (me)) +1
        if freeForks (Right (me))  = 2 then
            signal (ready (Right (me)))
        end if
        if freeForks (Left (me))  = 2 then
            signal (ready (Left (me)))
        end if
    end PutDown

initially
    imports (var freeForks)
    begin
        var j: 1..numberPhilos := 1
        loop
            freeForks (j) := 2
            exit when j = numberPhilos
            j:= j + 1
        end loop
    end

end {Forks} monitor

procedure CommonPhilosopher (i: 1..numberPhilos) =
    imports (var Forks)
    begin
        loop
            Forks.PickUp (i)
            Busy eating
            Forks.PutDown (i)
            Busy thinking
        end loop
    end CommonPhilosopher

process Philosopher1
    imports (CommonPhilosopher)
    begin
        CommonPhilosopher (1)
    end Philosopher1
```

```
process Philosopher2
    imports(CommonPhilosopher)
    begin
        CommonPhilosopher(2)
    end Philosopher2
```

...other philosophers...

end {Spaghetti} module

When a philosopher returns his forks in the PutDown entry, he should increase the fork counts of his neighbors. If the philosopher then finds that either (or both) of his neighbors has two forks available, he should signal the appropriate neighbor. The philosophers therefore pass the ability to access the forks among themselves using signal statements.

In our CE solution to the dining philosophers problem, each philosopher process calls a common procedure, supplying his number as the argument. This CE program illustrates several language features, among them an array of conditions (ready).

We now examine the solution for deadlock and indefinite postponement. Deadlock would occur if a philosopher became blocked and could not continue executing regardless of the future of the system. Let us look at the program to see where this might occur. A philosopher cannot become blocked forever when eating or thinking. Therefore, we look at execution in the two monitor entries to see if deadlock can occur there. The PickUp entry shows that a requesting philosopher suspends his execution when his two forks are not available. The philosopher does not pick up one fork and wait for the other. This means that the system can never get into the state where each philosopher holds one fork (his left one, say) and is waiting for the other -- this is deadlock, brought about by holding resources while requesting others. The rest of the PickUp entry shows that the fork counts are correctly decreased. The PutDown entry increases the fork counts and correctly signals other waiting philosophers.

The above discussion shows informally that deadlock cannot occur in our solution. Indefinite postponement is a different matter.

Indefinite postponement occurs if a philosopher becomes blocked and there exists a future execution sequence in which he will remain forever blocked. Consider the following situation for philosophers 1, 2, and 3 in which philosopher 2 is indefinitely postponed.

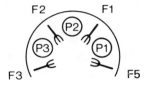

Suppose philosopher 1 picks up his two forks (forks 5 and 1). Next, philosopher 3 picks up his two forks (forks 2 and 3). Next, philosopher 2 enters PickUp and finds that his required forks are being used, so he waits. Next, the following unfortunate sequence occurs repeatedly. Philosopher 1 puts down his forks, thinks, and then picks them up again; then, philosopher 3 puts down his forks, thinks, and picks them up again, and so on. During this repeated sequence, at least one of the forks of philosopher 2 is always being used. Given that it is possible for this sequence to repeat indefinitely, we see that philosopher 2 can suffer indefinite postponement. We have not solved the problem!

There are several ways to overcome this defect. The most obvious one keeps track of the "age" of requests for forks, and when one request gets too old, other requests are held up until the oldest request can be satisfied. This could be done, for example, by counting the number of meals enjoyed by a philosopher's two neighbors while he is waiting for his forks. If he is bypassed more than, say, 10 times, his neighbors are blocked until he gets a chance to pick up his forks. (Does this solution allow maximum parallelism?) We leave this extension to the reader (see the Exercises at the end of this chapter).

This concludes our discussion of the dining philosophers problem. We saw that deadlock was avoided, but our simple solution unfortunately did not rule out indefinite postponement.

READERS AND WRITERS

We will now consider the problem of several processes concurrently reading and writing the same file. Any number of reader processes (processes accessing but not altering information) may access the file simultaneously. However, any writer process updating the file must have exclusive access to the file; otherwise, inconsistent data may result.

This problem arises in an airline reservations system, for example. Several ground personnel, each using a computer terminal, are issuing boarding passes for a flight. Reservations for the flight have been stored in a file. Reading the file allows an attendant to verify a passenger's reservation. Writing the file allows an attendant to add a new passenger to the flight when a customer arrives at the counter without a prior booking.

The problem, as stated so far, is an extension of the mutual exclusion example (Counting) given in Chapter 3. There still must be mutually exclusive access to the reservations file, but the new feature is that any number of reader processes may be accessing the file simultaneously. Writers still need exclusive access, however. The enqueue/dequeue feature discussed in Chapter 2 can solve the problem as stated.

A simple solution to this problem using monitors might be developed as follows. A reader wishing to access the file calls a monitor entry named StartRead. If there is no active writer (there may be several active readers), the number of active readers is increased by 1 and the new reader accesses the file. If there is an active writer, the reader waits. When a reader is waked up from a wait in StartRead, it signals other waiting readers that they also may access the file. A reader finishing accessing the file calls a monitor entry EndRead. It decreases the number of active readers by 1, and if this number reaches zero, it signals a writer that the file is available. (Readers cannot be waiting.) A writer wishing to access the file calls a monitor StartWrite. If there is an active writer or at least one active reader, the writer waits. A writer that is finished with the file calls EndWrite, and signals waiting readers or writers that the file is now available.

This simple solution unfortunately has an important defect. Once one reader begins accessing the file, the writers may be indefinitely postponed by a heavy stream of reader processes accessing the file. Some additional restriction on the problem must be made to remove the possibility of indefinite postponement.

We will impose the following requirements on the order of accessing the file. (They make the example more realistic and indefinite postponement is avoided.) We require that a new reader not be permitted to start if there is a writer waiting for the currently active readers to finish. Similarly, we require that all readers waiting at the end of a writer execution be given priority over the next writer. This latter restriction avoids the danger of the indefinite postponement of readers because of many active writers.

We now discuss a solution to the readers and writers problem with these restrictions. We will use a monitor with four entries (the names of these entries are the same as those above, but their structure is different): a reader process calls StartRead before reading and EndRead after reading; a writer process calls StartWrite before writing and EndWrite after writing. The following rules for the entries satisfy our ordering requirements.

StartRead. If there is an active writer or a waiting writer, the reader waits. When a waiting reader is awakened, it signals other readers to become active.

EndRead. If the finishing reader finds that it is the last active reader, it signals a waiting writer.

StartWrite. If there are active readers or if there is an active writer, the new writer waits.

EndWrite. If there are readers waiting, the finishing writer signals a reader. Otherwise, it signals another writer.

We will now consider these rules for these entries in more detail. In StartRead, a reader waits if either of two situations exists: there is an active writer (obviously) or there is a waiting writer. Because a reader in StartRead needs to distinguish between readers and writers in waiting to access the file, there should be separate conditions on which the readers (okToRead) and writers (okToWrite) wait. The empty pre-defined function can be used to test whether there are processes waiting on these conditions. Once a waiting reader in StartRead is resumed, it increases the number of active readers (numberReading) by one. This reader knows that the file is now available for reading, so it signals another reader that was waiting for access to the file. A resumed reader in StartRead thus contributes to a cascade of signaling readers which are waiting. No other process (such as an arriving reader or writer) can enter the monitor while this cascade is in progress. In time, all readers that were waiting for access after a writer are signaled. This version of StartRead differs from our previous version because now a reader waits if there is a waiting writer.

In EndRead, a reader decreases the number of active readers by one. If it finds that it is the last reader accessing the file, it signals a waiting writer. The logic of this entry is therefore identical to that of the previous version of EndRead.

In StartWrite, a writer waits if either of two situations exists: there is at least one active reader or there is an active writer. The Boolean variable activeWriter records whether writing is taking place. The logic of this entry is therefore identical to that of the previous version of StartWrite.

In EndWrite, a writer first checks if there are readers waiting to access the file. If there are, the writer signals a waiting reader. If there are not, the writer signals a waiting writer. This again illustrates the need for separate condition variables for readers and writers. EndWrite differs from our previous version because here a writer tries to signal a reader before it signals a writer.

Here is this solution to the readers and writers problem implemented in CE:

```
var ReadersAndWriters:
    module

        var FileAccess:
            monitor
                exports(StartRead, EndRead, StartWrite, EndWrite)

                var numberReading: UnsignedInt := 0
                var activeWriter: Boolean := false
                var okToRead: condition {When not activeWriter}
                var okToWrite: condition {When not activeWriter
                                and numberReading = 0}

                procedure StartRead=
                    imports(var numberReading, activeWriter,
                        var okToRead, okToWrite)
                    begin
                        if activeWriter or not empty(okToWrite) then
                            wait(okToRead)
                        end if
                        assert(not activeWriter)
                        numberReading := numberReading + 1
                        signal(okToRead) {Allow other readers in}
                    end StartRead

                procedure EndRead=
                    imports(var numberReading, var okToWrite)
                    begin
```

```
                numberReading := numberReading - 1
                if numberReading = 0 then
                    signal(okToWrite)
                end if
            end EndRead

        procedure StartWrite =
            imports(numberReading, var activeWriter,
                var okToWrite)
            begin
                if activeWriter or numberReading not = 0 then
                    wait(okToWrite)
                end if
                assert(not activeWriter and numberReading=0)
                activeWriter := true
            end StartWrite

        procedure EndWrite =
            imports(var activeWriter,
                var okToRead, var okToWrite)
            begin
                activeWriter := false
                if not empty(okToRead) then
                    signal(okToRead)
                else
                    signal(okToWrite)
                end if
            end EndWrite

    end {FileAccess} monitor

process Reader
    imports(var FileAccess, ...)
    begin
        loop
            ...
            FileAccess.StartRead
            Busy reading
            FileAccess.EndRead
            ...
        end loop
    end Reader
```

```
process Writer
    imports(var FileAccess, ...)
    begin
        loop
            ...
            FileAccess.StartWrite
            Busy writing
            FileAccess.EndWrite

            ...
        end loop
    end Writer
```

end {ReadersAndWriters} module

The restrictions of the readers and writers problem require that: (1) waiting readers are given priority over waiting writers after a writer finishes, and (2) a waiting writer is given priority over waiting readers after all readers finish. It is interesting to note that this form of "precedence" scheduling using signal statements is accomplished *without* priority conditions. Simple tests on numberReading, activeWriter, empty(okToRead) and empty(okToWrite) suffice to achieve the desired order of file access.

Finally, we discuss the signal statement in StartRead. A reader executing this signal suspends execution and allows a waiting reader to enter the monitor; this creates the cascade of resuming waiting readers. When the last waiting reader is resumed, it will signal an empty condition. The semantics of the signal statement in this situation are that the signaling process continues execution, possibly after other monitor entries are entered. The correctness of our solution is not affected, however. No writer can intervene during any of the suspended reader executions because the writer first checks on numberReading in the StartWrite entry. In these cases, numberReading is greater than 0, thanks to the increment of numberReading in StartRead before signaling okToRead.

A reader attempting to enter the monitor by a statement during the cascade of resuming readers will be blocked until the monitor becomes free. At that point, if the reader finds that there is a waiting writer, it will wait; if there is not a waiting writer, it will proceed. Although we have not shown it, the FileAccess monitor can support multiple readers and writers.

This concludes our discussion of the readers and writers problem. We saw that some complex scheduling decisions could be made without the need for priority conditions. The next example shows that priority conditions are sometimes convenient.

SCHEDULING DISKS

Good scheduling algorithms can greatly improve the performance of operating systems. For example, the average turnaround time for jobs can be minimized by running short jobs before long jobs.

In this section, we want to minimize the time that processes wait for disk input and output. Before discussing disk scheduling algorithms, we need to know how disks access their data. A disk consists of a collection of platters, each with a top and bottom surface, attached to a central spindle and rotating at constant speed. There is usually a single arm, with a set of read/write heads, one per surface; the arm moves in or out, across the disk surfaces. This type of disk is called a movable head disk. When the arm is at a given position, the data passing under all read/write heads on all platters constitutes a *cylinder*. At a given cylinder position, the data passing under a particular read/write head constitutes a *track*.

Files of data are stored on a disk. A file consists of records; a record consists of fields of data. Some disks allow many records on a track. On other disks, a record must correspond exactly to a track. A program requests a data transfer to or from a disk by giving the cylinder number, track number, and record number.

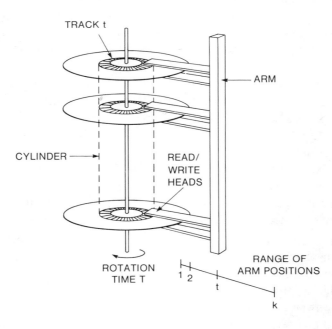

There are various delay factors associated with accessing data on a movable head disk. *Seek time* is the time needed to position the arm at the required cylinder. *Rotational delay* is the time needed for the disk to rotate so that the desired record is under the read/write head. *Transmission time* is the time needed to transfer data between the disk and main memory.

Seek time increases with the number of cylinders that the arm is moved. Rotational delay can vary from zero to the time needed to complete one revolution; on the average, it is one-half the rotation time. Transmission time is dependent on the rotation time and recording density, and these factors vary from disk to disk. The following table shows that the seek time delay is the dominant factor in a typical transfer of a 1000-character record on a movable head disk.

Factor	*Length of Time*
Seek time	30.0 ms (milliseconds)
Rotational delay	8.3 ms
Transmission time	1.2 ms

Generally, a simple scheduling algorithm can do nothing to decrease rotational delay or transmission time. But ordering of disk read/write requests can decrease the average seek time. This can be accomplished by favoring those requests that do not require much arm motion.

Perhaps the simplest scheduling algorithm to implement is the FIFO algorithm; it moves the disk arm to the cylinder with the oldest pending request. With light disk traffic, the FIFO algorithm does a good job. But when the queue of disk requests starts building up, the simple nature of FIFO results in unnecessary arm motion. For example, suppose a sequence of disk requests arrives to read cylinders 12, 82, 12, 82, 12, and 82, in that order. (Such a sequence can easily occur if several jobs are using a file on cylinder 12 while others are using a file on cylinder 82.) The FIFO algorithm will unfortunately cause the disk to seek back and forth from cylinder 12 to 82 several times. A clever scheduling algorithm could minimize arm motion by servicing all the requests for cylinder 12 and then all the requests for cylinder 82. We will discuss two scheduling algorithms that are more sophisticated than FIFO.

One such algorithm is the shortest seek time first (SSTF) algorithm. It operates as follows: the request chosen to be serviced next is the one that will move the disk arm the shortest distance from its current cylinder. The SSTF algorithm thus attempts to reduce disk arm motion. However, it can exhibit unwanted behavior; requests for certain cylinder regions on the disk may be indefinitely overtaken by requests for other cylinder regions closer to the current disk arm position. This results in certain disk requests not being serviced at all.

The SCAN algorithm is another scheduling algorithm that achieves short waiting times but prevents indefinite postponement. It attempts to reduce excessive disk arm motion and average waiting time by minimizing the frequency of change of direction of the disk arm. The SCAN algorithm operates as follows: while there remain requests in the current direction, the disk arm continues to move in that direction, servicing the request(s) at the nearest cylinder; if there are no pending requests in that direction (possibly because an edge of the disk surface has been encountered), the arm direction changes, and the disk arm begins its sweep across the surfaces in the opposite direction. This algorithm has also been called the elevator algorithm because of the analogy to the operation of an elevator, that runs up and down, receiving and discharging passengers.

A DISK ARM SCHEDULER

We will now give an implementation of the SCAN disk scheduling algorithm. SCAN gives good performance and illustrates the use of priority conditions. We will assume that there is a single disk and access to it is controlled by a monitor. The monitor has two entries:

Acquire(destCyl). Called by a process prior to transferring data to or from cylinder destCyl.

Release. Called by a process after it has completed data transfer on the current cylinder.

The monitor must guarantee that only one process at a time uses the disk. The local data of the monitor keeps track of the current state of the disk. There must be a current disk arm position, armPosition (which can vary from 0 to the maximum cylinder number, maxCylNumber), the current direction of the arm sweep, direction (up or down), and a Boolean variable called inUse indicating whether the disk is currently busy. The up direction corresponds to increasing cylinder numbers, the down direction to decreasing cylinder numbers. A process using the disk executes the following:

> Acquire(destination cylinder)
> Read from or write to disk
> Release

Here is the monitor implementing the SCAN algorithm:

```
1   const maxCylNumber := 400

2   var Scan: {Disk scheduling by SCAN (elevator) algorithm}
```

```
3      monitor
4          imports(maxCylNumber)
5          exports(Acquire, Release)

6          var armPosition: 0..maxCylNumber := 0
7          var inUse: Boolean := false
8          pervasive const down := 0
9          pervasive const up := 1
10         var direction: down..up := up
11         var downSweep: priority condition
               {When not inUse and direction=down}
12         var upSweep: priority condition
               {When not inUse and direction=up}

13         procedure Acquire(destCyl: 0..maxCylNumber) =
14             imports(maxCylNumber, var armPosition, var InUse,
15                 direction, var downSweep, var upSweep)
16             begin
17                 if inUse then
18                     if armPosition < destCyl or
                           (armPosition=destCyl and direction=down) then
19                         wait(upSweep, destCyl)
20                         assert(not inUse and direction=up)
21                     else
22                         wait(downSweep, maxCylNumber-destCyl)
23                         assert(not inUse and direction=down)
24                     end if
25                 end if
26                 inUse := true
27                 armPosition := destCyl {Record arm position}
28             end Acquire

29         procedure Release=
30             imports(var inUse,
31                 var direction, var downSweep, var upSweep)
32             begin
33                 inUse := false
34                 if direction = up then
35                     if empty(upSweep) then
36                         direction := down
37                         signal(downSweep)
38                     else
```

```
39                       signal(upSweep)
40                   end if
41               else
42                   if empty(downSweep) then
43                       direction := up
44                       signal(upSweep)
45                   else
46                       signal(downSweep)
47                   end if
48               end if
49           end Release

50   end {Scan} monitor
```

It is particularly enlightening to discuss the SCAN monitor in detail. In the Acquire entry, if the disk is free the arm is moved to the desired cylinder and the process leaves the monitor. Otherwise, the process waits. It is not sufficient to use a single condition for waiting. (Why?) What we need is a set of conditions that relate the priority of a waiting request to the distance the requested cylinder is from the current cylinder. We could use an array of conditions, one for each cylinder. Then, when a process releases the disk, it would determine the "closest" non-empty condition waiting list in the current arm direction and signal that condition. This approach is clumsy and we will use priority conditions instead.

There are two priority conditions, each corresponding to a given arm direction (upSweep or downSweep). In the Acquire entry, a process waits on upSweep if its destination cylinder has a larger number than the current cylinder. A process waits on downSweep if its destination cylinder has a lower number than the current cylinder. Two questions arise: What happens if the destination cylinder equals the current cylinder? What priorities are specified in these waits?

We answer the priority question first. The priorities must indicate the distance the destination cylinder is from one end of the disk. For example, suppose the current arm position is at cylinder 25 and the direction is up. What happens with processes that request cylinders 100 and 200? Both requests are ahead of the disk arm and therefore will wait on upSweep. The request for cylinder 100 is closer than the request for cylinder 200, so cylinder 100 has priority over cylinder 200 (and therefore has a lower priority number). In our example, the request for cylinder 100 has a priority value of 100, while the one for cylinder 200 has a priority value of 200. Cylinder 100 will therefore be serviced before cylinder 200 on the upsweep.

Consider another example. Suppose the current cylinder is 250, and the direction is up. What happens with processes that request cylinders 150 and 50? Both requests are behind the disk arm and therefore will wait on downSweep. When the disk arm begins its sweep in the down direction, cylinder 150 is closer to it than cylinder 50. Assuming maxCylNumber is 400, cylinder 150 has priority over cylinder 50 (and should have a lower priority number). This can be accomplished by subtracting the destination cylinder from the maximum cylinder number to produce the priority value; the result will always be in the range from zero to maxCylNumber and will indicate the relative distance of arm motion. Cylinder 150 will have a priority value of 250 and cylinder 50 will have a priority value of 350. Cylinder 150 will therefore be serviced before cylinder 50 on the down sweep.

In our example, the priority values are specified in the wait statements in lines 19 and 22. We have been careful in our discussion here to be explicit about the arm direction, because the correct condition to wait on and the correct priority depend on that direction. But the arm direction is only tested (in line 18) when the destination cylinder equals the current cylinder.

The answer to this apparent mystery is that the arm direction generally does not matter. If the arm position is less than the destination cylinder, the process should wait on upSweep regardless of the arm direction. If the arm position is greater than the destination cylinder, the process should wait on downSweep regardless of the arm direction. Only when the arm position equals the destination cylinder does the arm direction matter. We are back to the first question we asked, so it is appropriate to answer it now.

To see what happens if the destination cylinder equals the current cylinder, we will discuss an alternative to line 18, showing that it can lead to indefinite postponement. (Questions 10 and 11 in the Exercises mention other alternatives.)

Consider the following alternative to line 18:

if armPosition < destCyl then

A process with a destination cylinder equal to the current cylinder will wait on downSweep in line 22. Suppose the current direction of the disk is down. A process releasing the disk will signal downSweep in line 46 because there is a process waiting on it. If there is a stream of processes similar to the first one in this example, all with requests for the current cylinder, the disk arm will remain at the current cylinder servicing all these requests. Indefinite postponement results because requests for other cylinders are being continually denied.

The alternatives here and in the Exercises show that implementing the SCAN algorithm is a tricky matter. How does the current line 18 preclude indefinite postponement?

if armPosition < destCyl or
    (armPosition = destCyl and direction = down) then

A process with a destination cylinder equal to the current cylinder will wait on upSweep when the direction is down. When the direction is up, such a process will wait on downSweep. Thus, in the original line 18, a process with a destination cylinder equal to the current cylinder waits on the condition variable associated with the opposite of the current arm direction. Such a request does not cause the disk arm to remain at the current cylinder because it does not wait on the condition associated with that direction. The request will not be serviced on the current sweep, but on the next sweep in the opposite direction. Indefinite postponement is precluded in this way.

There is another benefit in the way we organized our solution. All requests waiting for a cylinder before the arm gets to that cylinder will be served on the same sweep. This benefit comes from the priority values. All requests for the same cylinder have the same priority value. When the destination cylinder equals the current cylinder, the priority value associated with the destination cylinder is the smallest priority value in the current arm direction. Thus, the signals in lines 39 and 46 will resume all processes that requested the current cylinder and were waiting before the arm got to that cylinder, before going on to another cylinder in the same direction or changing directions. This improves performance because it reduces waiting time.

In the Release entry, control of the disk is returned and the process to be serviced next is signaled. If there is a waiting request in the current direction, it is signaled (lines 39 and 46). If there is no waiting request in the current direction, the arm direction is changed (lines 36 and 43) and a waiting request in the new direction is signaled (lines 37 and 44). In all cases, the process to be serviced next is the one that has the closest request in the current arm direction.

We have devoted much space to considerations of indefinite postponement. Deadlock is a simpler matter.

Deadlock cannot occur in our solution. Processes are never blocked in their process code. In the monitor code, only a single resource (the disk) is being used; processes can never hold some resource units while requesting others. Finally, processes wait on the correct conditions and specify the correct priority values.

This concludes our discussion of the SCAN disk scheduling algorithm. Priority conditions were used to advantage for scheduling. We showed that indefinite postponement was prevented in our solution.

BUFFER ALLOCATION FOR LARGE MESSAGES

In this section we discuss an example that builds upon the circular buffer manager (MailBox) of Chapter 4. In that example, a producer wishing to add information to the queue called the monitor entry Send. The buffers were represented by an array, and the information to be added was stored in an array element. A consumer wishing to remove information from the queue of buffers called the monitor entry Receive.

That example was suitable for the case of low-volume information (e.g., message queues), where the copying of information into and out of the buffers did not seriously degrade process performance. Such a design is not suitable for the transmission of high-volume information (e.g., files of records) because of the large amount of data movement. For large messages, we can use a MailBox monitor to store the *locations* of the buffers of information, rather than the actual information. Data movement will be reduced because only locations will be added to or removed from the queue.

We now describe the large message or "pipeline" environment in more detail. There are several pairs of producers and consumers, each producer generating data for its consumer through a queue unique to the pair. Shared among all the producer/consumer pairs is a pool of free buffers. (The shared pool allows more efficient use of buffers.) A producer repeatedly acquires a free buffer (and notes its buffer location), fills the buffer, and adds the buffer location to the queue shared with its consumer. A consumer repeatedly removes a buffer from the queue, empties the buffer, and releases the buffer (by returning its buffer location to the pool of free buffer locations).

A producer and consumer need mutually exclusive access to their common queue, so accesses to the queue will go through a monitor. We can use the MailBox monitor from Chapter 4; each producer/consumer pair will have its own MailBox monitor, with entries Send and Receive. Producers and consumers must go through a monitor (named BufferManager) to Acquire and Release buffer locations.

The code for the producer and consumer processes appears below. Send and Receive are entries of a MailBox monitor local to a producer/consumer pair. Acquire and Release are entries of BufferManager common to all processes. The BufferManager allocates buffer locations from the pool and returns them to the pool.

```
process Producer
    imports(var MailBox, var BufferManager, var buffers)
    begin
        var bufferNumber: BufferIndex
        loop
            BufferManager.Acquire(bufferNumber)
            Fill buffer(bufferNumber)
            MailBox.Send(bufferNumber)
        end loop
    end Producer

process Consumer
    imports(var MailBox, var BufferManager, var buffers)
    begin
        var bufferNumber: BufferIndex
        loop
            MailBox.Receive(bufferNumber)
            Empty buffer(bufferNumber)
            BufferManager.Release(bufferNumber)
        end loop
    end Consumer
```

The BufferManager follows next. It contains the Acquire and Release entries which the Producer and Consumer processes, given above, call. The list of buffer locations is kept in an array named pool. This array is managed as a stack, which uses top to point to the next free buffer location. When a producer finds that top=0, the free list has been exhausted and the producer must wait (on the condition bufferFree). Note the way in which the buffer pool is initialized.

```
1   pervasive const poolSize := 100
2   pervasive type BufferIndex = 1..poolSize

3   var BufferManager:
4       monitor
5           exports(Acquire, Release)

6           var pool: array BufferIndex of BufferIndex

7           var top: 0..poolSize := poolSize
8           var bufferFree: condition {When top > 0}
```

```
9            procedure Acquire(var buffLoc: BufferIndex) =
10               imports(pool, var top, var bufferFree)
11               begin
12                  if top = 0 then
13                      wait(bufferFree)
14                  end if
15                  buffLoc := pool(top)
16                  top := top - 1
17               end Acquire

18           procedure Release(buffLoc: BufferIndex) =
19               imports(var pool, var top, var bufferFree)
20               begin
21                  top := top + 1
22                  pool(top) := buffLoc
23                  signal(bufferFree)
24               end Release

25           initially
26               imports(var pool)
27               begin
28                  var i: BufferIndex := 1
29                  loop
30                      pool(i) := i
31                      exit when i = poolSize
32                      i := i + 1
33                  end loop
34               end

35       end {BufferManager} monitor
```

If the pool of free buffers is empty and several producer/consumer pairs are operating at widely different speeds, the scheduling policy in the BufferManager can degrade the performance of the process pairs. The typical FIFO policy of CE conditions will allocate alternate buffers to two competing producers; this seems reasonable at first glance. But if two competing consumers are a 2000 line/minute line printer and a 15 line/minute console typewriter, all buffers will eventually be allocated to the pair having the slower consumer (the console). The pair having the faster consumer will be reduced to the speed of the slower pair. (We do not consider pairs in which the consumer is always faster than the producer, because such pairs will only have a small number of buffers allocated to them.)

Under heavy load conditions, the pool of free buffers should be shared among the producer/consumer pairs in a reasonable manner. One scheduling policy that achieves a compromise between fast and slow pairs of processes is to allocate a free buffer location to the producer whose pair currently has the smallest number of buffers allocated to it. This method tries to keep a balanced system of competing pairs, operating far away from the undesirable situation of too many buffers committed to slow consumers.

This scheduling policy can be implemented using priority conditions. Two additional items are needed in the BufferManager monitor: a new parameter to Acquire, named pair, giving the pair number of the requesting process and a tally of the number of the buffers currently allocated to a pair, count(pair). Count(pair) is increased by one in the Acquire entry and is decreased by one in the Release entry. The priority wait appears in lines 12 through 14 as:

```
if top = 0 then
    wait(bufferFree, count(pair))
end if
```

The signal statement in the Release entry will activate the process having the smallest priority value; this allocates a free buffer to the pair currently having the smallest number of buffers allocated to it.

This completes our discussion of the buffer allocator for large messages. This allocator uses the circular buffer monitor of Chapter 4 to transmit large amounts of information; it avoids the overhead of copying by passing locations of data rather than the data itself.

CHAPTER 5 SUMMARY

In this chapter we stated four problems in concurrent programming and gave solutions to them in CE.

The dining philosophers problem involved the concurrent accessing of a set of common resources. A naive solution of picking up one fork and then the other fork can lead to deadlock. A solution of picking up both forks at once cannot lead to deadlock. Our solution avoided deadlock but did not avoid indefinite postponement (starvation). We showed an execution sequence in which a philosopher could be indefinitely overtaken by his neighbors.

The readers and writers problem involved the concurrent reading and writing of a file. Without putting additional restrictions on the simple problem, the indefinite postponement of writers was a possibility. We imposed restrictions that specified the desired order of accessing the file. Deadlock and indefinite postponement were not possible in our solution. The complex scheduling rules followed after a signal statement were implemented using two conditions, one for readers and one for writers.

The disk scheduling problem involved the priority ordering of accesses to a movable head disk. Following a discussion of disks and disk scheduling, we developed a SCAN algorithm. We showed how priority conditions were used to service the next closest request in the current arm direction. We also showed that several ways of ordering pending requests lead to indefinite postponement. Our solution avoided indefinite postponement and deadlock, and gives good performance to batches of requests for the same cylinder.

The large message problem involved the transmission of high volume information without the overhead of data movement. We used a circular buffer manager to store the locations of the buffers of information, rather than the actual information. The locations were acquired from and released to a pool of free buffer locations, controlled by a monitor. When pairs of producers and consumers operate at widely different speeds, our initial solution could produce undesirable performance. We amended the solution by introducing priority waits to achieve improved resource allocation.

CHAPTER 5 BIBLIOGRAPHY

Our version of the dining philosophers problem and solution is based on material from Brinch Hansen [1973] and Kaubisch et al [1976]. The readers and writers problem and the disk arm scheduler come from Hoare [1974]. An early version of the readers and writers problem was given by Courtois et al [1971]. Material on numerical and simulation studies of disk scheduling can be found in Teorey and Pinkerton [1972]. The large message problem was first studied by Dijkstra [1972]. Further comments on it were made by Brinch Hansen [1973] and Hoare [1974]. Dijkstra's article [1968] remains a good treatment of concurrency issues, although monitors are not used.

Brinch Hansen, P. *Operating Systems Principles.* Prentice-Hall (1973).

Brinch Hansen, P. Concurrent programming concepts. *Computing Surveys* 5,4 (December 1973), 223-245.

Courtois, P.J., Heymans, F., and Parnas, D.L. Concurrent control with readers and writers. *Comm. ACM 14,*10 (October 1971), 667-668.

Dijkstra, E.W. Cooperating sequential processes. In *Programming Languages* (F. Genuys, editor), Academic Press (1968).

Dijkstra, E.W. Information streams sharing a finite buffer. *Information Processing Letters 1*,5 (October 1972), 179-180.

Hoare, C.A.R. Monitors: an operating system structuring concept. *Comm. ACM 17*,10 (October 1974), 549-557.

Kaubisch, W.H., Perrott, R.H., and Hoare, C.A.R. Quasiparallel programming. *Software-Practice and Experience*, Vol. (1976), 341-356.

Teorey, T.J. and Pinkerton, T.B. A comparative analysis of disk scheduling policies. *Comm. ACM 15*,3 (March 1972), 177-184.

CHAPTER 5 EXERCISES

1. Implement and run all examples given in this chapter in CE. Insert output statements at appropriate places in the examples to show the progress of the processes and the states of the resources.

Questions 2-7 refer to the dining philosophers problem.

2. Discuss the effects of changing the PutDown entry to have the following form: increase left count, if left philosopher has two forks then signal him, increase right count, if right philosopher has two forks then signal him.

3. Discuss the following proposed solution: a hungry philosopher first attempts to pick up his left fork through a monitor call; he then attempts to pick up his right fork through another monitor call; holding both forks, he begins eating.

4. Discuss the following proposed solution: a hungry philosopher first attempts to pick up his left fork through a monitor call; he then attempts to pick up his right fork through another monitor call; if the right fork is available, he picks it up and begins eating; otherwise, he puts down his left fork and repeats his cycle.

5. Discuss the following proposed solution: all philosophers are initially thinking; each philosopher waits until both his neighbors are thinking; he then stops thinking, picks up both his forks, and starts eating; when finished eating, he puts down the forks and starts thinking.

6. Discuss the following proposed solution:

       if right fork is taken then
           wait for right fork

```
         if left fork is taken then
             wait for left fork
         end if
    else
         if left fork is taken then
             wait for left fork
         elseif right fork is taken then
             wait for right fork
         end if
    end if
```

7. [S.S. Toscani] What happens if the following rule is observed by the philosophers: all philosophers, except one that is unconventional, acquire first the left fork and then the right. The unconventional one does the opposite, acquiring the right and then the left. Can deadlock occur? Can indefinite postponement occur?

Questions 8-9 refer to the readers and writers problem.

8. Discuss the solution proposed by Kaubisch et al [1976]. In particular, compare their StartRead entry with our StartRead entry and their use of a counting variable for the number of writers with our use of a Boolean variable.

9. Courtois et al. [1971] considered two variants of the simple readers and writers problem. These variants differed from our version in the restrictions they imposed on the order of accessing the file.

    a) No reader should be kept waiting unless a writer has already obtained permission to use the file. That is, no reader should wait simply because a writer is waiting for other readers to finish.

    b) Once a writer is ready to write, it performs its write as soon as possible. That is, no writer should wait simply because there is a stream of reader requests waiting after an active writer.

Why is a solution to problem b) not a solution to problem a)? Show that it is possible for a writer to be indefinitely postponed in problem a). Show that it is possible for a reader to be indefinitely postponed in problem b). Develop solutions to these variants in CE and compare them to the solution given in this chapter.

Questions 10-14 refer to the disk arm scheduling problem.

10. Show that the following alternative to line 18 in the SCAN monitor allows indefinite postponement.

    if armPosition $<=$ destCyl then

11. Show that the following alternative to line 18 in the SCAN monitor allows indefinite postponement.

> if armPosition < destCyl or
> (armPosition = destCyl and direction=up) then

12. Implement the FIFO disk scheduling algorithm in CE and test it using a sequence of disk requests.

13. Implement the SSTF disk scheduling algorithm in CE and test it using a sequence of disk requests.

14. A disk can be simulated by a process that executes the following statements.

```
loop
    GetCylToSeek(destCyl)
    if destCyl > currentCyl then
        distance := destCyl - currentCyl
    else
        distance := currentCyl - destCyl
    end if
    busy(distance)
end loop
```

Compare the average waiting time for disk I/O using the SCAN, SSTF, and FIFO algorithms. Note: the simulation (busy) feature of CE provides utilization statistics automatically.

Questions 15-16 refer to the buffer allocator for large messages.

15. In the large message problem, what steps can be taken to overcome the difficulties caused by a consumer stopping altogether.

16. Discuss methods different from that in the text for achieving reasonable resource allocation in a heavy-load condition.

Questions 17-19 are new problems.

17. Develop a simulation of the sleeping barber problem [Dijkstra 1968] in CE: There is a barbershop with two rooms, one with the barber's chair, the other a waiting room. Customers enter from the outside into the waiting room one at a time; from the waiting room, they can proceed into the barber's room. The entrances to the two rooms are side-by-side and share a sliding door (which always closes one of them). When the barber finishes, the customer leaves by a separate exit and the barber inspects the waiting room by opening the door to it. If the waiting room is not empty, he invites the next customer in; otherwise, he goes to sleep in one of the waiting room chairs. When an entering customer finds a sleeping barber, he wakes up the barber; otherwise, he waits his turn.

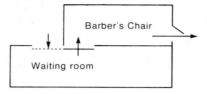

18. Develop a solution to the alarm clock problem in CE and discuss its overhead: A program wishes to delay itself a specified number of time units, or "clock ticks". Assume that a hardware clock can update a simulated software clock which is available for program inspection. The hardware clock can be simulated by a process that executes the following:

```
loop
    tick
    busy(1)
end loop
```

19. "Spoons" is a card game that is played by a group of N people sitting in a circle around N-1 spoons. Initially each player is dealt four cards. A player alternately discards to a pile to his left and draws from a pile to his right. When his hand has four of a kind, he grabs a spoon. Whenever a spoon is grabbed, the remaining players each attempt to grab a spoon. When the dust settles, the loser is the person who did not get a spoon. (In an actual game, this is repeated, each round eliminating one person until only the winner remains.)

Program a simulation of Spoons in CE. Each process (player) executes the following program:

```
loop
    exit when (hand contains four of a kind) or
        (someone has grabbed a spoon)
    Discard a card to the left-hand pile
    Pick up a card from the right-hand pile
end loop
Attempt to grab a spoon
Loser := (Did not get a spoon)
```

You may wish to use a deck of 24 cards, split into 6 different kinds, four cards of each kind. Assume four players, each holding four cards. Between each pair of players is a pile that initially holds two cards.

Chapter 6

UNIX:
USER INTERFACE
AND FILE SYSTEM

In the preceding chapters we have discussed programming techniques and the Concurrent Euclid language with an emphasis on concurrency. These techniques are useful in the construction of operating systems, real time systems, and special purpose microprocessor systems. The rest of the book is concerned with the design and implementation of operating systems, concentrating on Unix. The Unix operating system is being used on thousands of computer systems. Because of its widespread acceptance and its simple but powerful facilities, it is an excellent example to study in order to learn about operating systems.

This chapter introduces Unix's interface to interactive users and discusses the organization of its file system. Readers who are users of Unix will be familiar with much of this material. The following two chapters discuss the Unix system calls to control user processes and the major data structures in the Unix nucleus. Following these is a chapter giving the design of Tunis, which is a compatible replacement for the Unix nucleus written in CE.

HISTORY AND OVERVIEW OF UNIX

Unix was designed and implemented by two people, D.M. Ritchie and K. Thompson, at Bell Laboratories during 1969-71. It is a medium to small scale operating system. It works well on computer configurations having as little as 128K bytes of main memory and typically occupies about 60K bytes of such a system. It has proven to be a practical system that is particularly good at supporting text manipulation and program development.

Although not originally designed for portability, Unix has proved to be relatively easy to move to new computer architectures. It was first widely used on the Digital Equipment PDP-11. It has been ported to various architectures including the Interdata 8/32, the Digital Equipment VAX and the Zilog Z8000.

Unix is flexible and convenient to use, especially for the sophisticated user. Its file system and command language allow the user to easily solve problems which were painful to handle in many predecessor operating systems.

It has a rather small and secure nucleus. In Unix terminology, this nucleus is called a "kernel". Unfortunately, this is at odds with the present book, which reserves the word "kernel" to mean a small module (e.g., 2K bytes) that does little more than handle interrupts and share CPU time among processes. When you read other material on Unix, be warned that what we call the Unix nucleus is elsewhere called the Unix kernel.

Unix is a process-based system meaning that all activities outside the nucleus run as processes. One of the main purposes of the nucleus is to implement multiprogramming to support these processes.

Unix was originally implemented in assembly language. It was later re-written in the programming language C. Several functions were added to Unix during this re-write. As a result of the re-write and the new functions, Unix grew in size by about one third. This is a modest price to pay for the resultant cleanliness, modifiability and portability gained by using a higher level language. As was mentioned in the chapter introducing Concurrent Euclid, C lacks many of the constructs of other high level languages. For example, it does not support array subscript checking or tight constraints on pointer use. It is because of these shortcomings that this book uses Concurrent Euclid instead of C.

Unix provides a clean, flexible interface to the user, and this interface is available on various different computers because Unix is relatively portable. This provides a great advantage to the user: he can easily move his skills, programs and data among different computer systems without learning new command languages and system conventions. This is analogous to the great advance that Fortran provided in the 1950's, in providing a convenient-to-use notation that could be used on different manufacturers' machines. By analogy, we are tempted to call Unix the "Fortran of operating systems." Unix is apparently destined to become the standard operating system for various application areas.

TYPICAL CONFIGURATIONS

Unix was first widely used at Bell Laboratories and at universities. Inevitably, there have evolved various, somewhat incompatible versions of Unix. Among the most widely used outside Bell Laboratories have been Version 6, Version 7 and Berkeley Unix; Berkeley Unix runs on the Digital Equipment VAX. A newer version of Unix, called System III, is expected to become widespread. For Versions 6 and 7, typical hardware configurations for Unix include:

CPU (PDP-11, Z8000, or similar processor)
256K bytes of main memory
10 CRT terminals
4 30-megabyte disk drives
1 tape drive (used for file backups and transporting
    programs and data among installations)

With the continuing drop in hardware costs, we will see more main memory and disk memory per terminal, and more single-user Unix systems.

MAJOR LAYERS OF UNIX

Unix is constructed of two major software layers. The lower layer is the nucleus. The higher layer consists of user processes, which are supported by the nucleus. We can consider that the people who use Unix form a layer outside the user process layer. The hardware, especially the disk drives, can be considered as another layer, which is lower than and supports the nucleus. Proceeding from outside to inside, these layers are:

(1) *Interactive users.* These are people typing at terminals. A user types *commands* to Unix's command processor, which is called the *shell.* It is called the shell because from the interactive user's point of view, it surrounds the rest of Unix and gives access to it, while from the system's point of view it surrounds the interactive user, handling interactions with him. There are other programs, such as Unix's editor, that also accept user's commands.

(2) *User processes.* These communicate with interactive users and invoke the nucleus. Programs such as the shell and the editor execute as user processes. Some so-called user processes carry out system functions such as creating file directories and listing files on the printer. Each user process has its own *virtual memory* or *address space* that does not overlap other processes or the nucleus. The only way a user process can communicate with another user process or the nucleus is via a system call (trap) to the

nucleus. The nucleus swaps or pages inactive user processes out to disk.

(3) *The nucleus.* This implements user processes and a disk resident file system. It carries out requests (system calls) by user processes. It handles interrupts and controls peripheral devices. The Tunis implementation of the nucleus is discussed in a later chapter.

(4) *Disks and peripheral devices.* The disks contain data structures that represent users' files. Access to other peripherals is supported by the nucleus, which makes them behave like "special" files.

SYSTEMS THAT ARE UNIX-COMPATIBLE

The interfaces to the two major software layers of Unix can be supported by non-Unix software. A system that supports shell commands, editor commands, etc. is *Unix-compatible at the command interface.* Operating systems that are quite different from Unix can support this interface. This is done by providing a program that behaves like the shell, another that behaves like the editor, and so on for other standard Unix programs.

A computer system that supports the same system calls as Unix's nucleus is *Unix-compatible at the system call interface.* Such a system allows user processes to execute as if they are running under a true Unix nucleus. For example, there is a software package called Eunice which approximately supports this interface under Digital Equipment's VMS operating system. A different approach is to replace the Unix nucleus from the ground up with an equivalent nucleus supporting Unix's system calls. This is done by the Tunis nucleus, which is described in a following chapter. Tunis is also *Unix-compatible at the disk interface*, meaning that disk packs containing files can be used interchangeably by Unix and Tunis.

In the rest of this chapter, we overview the command interface of Unix and the Unix file system.

LOGGING IN AND SIMPLE COMMANDS

When an interactive user wants to use Unix, he types his *account name* and *password*, as in this example:

```
login: rch
passwd: secret
Any news printed at login time
%
```

Unix types "login:" and the user types his account name, "rch" in this example. Then Unix types "passwd:" and the user types his password. The password, which is "secret" here, is not actually printed. Next, Unix may print various news, followed by a prompt character, which is shown as "%" here. The prompt character means that the shell is ready to accept commands from the interactive user. In the next chapter, we will explain how this protocol is implemented via process creation and the "exec" system call.

Unix maintains a file of the names of accounts (user IDs) along with an encryption of their respective passwords. It uses this file to verify that the login is for a legitimate account and password.

Once a user has logged in, he can type commands to the shell, such as

    % who

The who command lists the set of users currently logged in, for example, the following might be printed:

mckenzie	ttyd	Feb 28	12:56
jrc	ttyp	Feb 28	14:13
rch	ttyi	Feb 28	14:23

From this we can tell that users with the accounts mckenzie, jrc and rch are presently logged in; we are also told the times of their logins.

As an example of another command, here is the way to put a message on another user's terminal:

    % write jrc
    Don't forget our squash game at 16:00
    <control D>
    %

This causes the line about the squash game to be printed on jrc's terminal. The <control D> is a character that specifies the end of the message. Although we will not go into details here, it is possible for two users to hold a (clumsy) conversation by writing on each other's terminals. The percent signs are again prompt characters printed by the shell.

CREATING, LISTING AND DELETING FILES

The usual way of creating a text (ASCII) file under Unix is by using an editor. A number of different full screen editors are used under Unix. The original Unix editor, called ed, is line oriented. Here is a session using a version of ed to create a Concurrent Euclid program.

    % ed copy.e
    ?copy.e
    *a

```
{This is a CE program that reads and prints
    characters up to a period}
var copy:
    module
        ... body of module ...
    end module
```

```
*w
*q
%
```

The editor is invoked by typing its name, ed, and the name of the file to be edited, copy.e. The editor types "?copy.e" meaning that there is no existing file by that name, but that the file will be created. The editor's prompt character is "*", which is printed in the next line. Some versions of the editor do not use a prompt character and some differ in the information they print.

The user types "a" following the "*" to signify that he wants to append to the file. He then types the contents of the new file followed by a single period on a line, signifying the end of the text. The editor provides facilities for modifying the text, but we will not be going into these. Following the next editor prompt "*", the user types "w" meaning to write the text into the file copy.e. After the next prompt "*", he types q for quit; this terminates the editing session and the shell again types its prompt "%".

There are various ways of listing a file under Unix, the simplest being:

```
% cat copy.e
```

The name "cat" is short for catenate and is a misnomer for this use, which simply lists the text in copy.e on the screen. We can compile copy.e, as follows:

```
% cec copy.e
```

Assuming copy.e contains no compile time errors, this command creates a file named copy.out, which can be executed by:

```
% copy.out
```

The Concurrent Euclid program will now begin executing. Assuming this program is similar to the copying program presented in Chapter 3, it reads and writes a character by the statements:

```
IO.GetChar(ch)
IO.PutChar(ch)
```

and halts when ch is a period. So if we type:

> This is the end.

the program will read this line and have it printed on the screen (again):

> This is the end.

Next, the shell prints "%". Under Unix, a complete line is read from the terminal before giving any characters to the executing program (copy.out). That is why all of the first copy of "This is the end." is printed before its copy is printed.

If we want to list a copy of copy.e on the printer, we run a program such as lpr:

> % lpr copy.e

where lpr means "line printer".

When we have no more use for copy.e, we can remove (rm) it by:

> % rm copy.e

This destroys the file and recovers the disk space allocated to it.

THE DIRECTORY HIERARCHY

The files of Unix are arranged into a tree or hierarchic structure of directories. Each user has his own "home" directory in the tree. When the user logs on, he has immediate access to files in his own directory. If he wants to create a set of related files, he can create a sub-directory (a new node in the directory tree) to hold these files. For example, suppose that an author is a user of a Unix system and his account name is rch. His home directory (as well as his account) is named rch. Suppose he is writing a book, so he has created a sub-directory called book. In the book directory are individual files named ch1, ch2, etc. containing chapters of the book. (The sentence you are presently reading is located in file ch6 of R. C. Holt's book directory.) Here is an example directory structure for a Unix system.

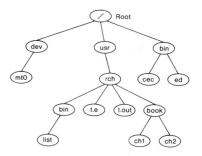

Each user, such as rch, has his own home directory in the tree of directories. In this example, the directory for rch contains four entries: bin, t.e, t.out and book. The bin and book nodes are directories while t.e and t.out are ordinary user files. Since t.e has the suffix ".e", we can assume that it is a Concurrent Euclid program.

If the book directory and its contents do not yet exist, rch can create them by:

    % mkdir book
    % cd book
    % ed ch1
    ...etc...

The mkdir (make directory) command creates the book directory, which initially has no entries. To move from the rch directory to the book directory, we use the cd (change directory) command. We create the file ch1 in the book directory using the Unix editor ed.

In the directory tree, rch is a sub-directory in the usr (user) directory. In turn, the usr directory is a sub-directory of the root of the directory tree, which has the name / (slash). Each file or directory has a *path name* that starts at the root; the path name includes the sub-directories down to the file. For example, the path name for ch1 is

    /usr/rch/book/ch1

The path name for the rch directory is

/usr/rch

When a path name begins with slash, this means the path starts at the root directory. If the path starts with a non-slash, then the path starts in the currently active directory. For example, when user rch logs into Unix, his active directory is rch and the path name for ch1 will be book/ch1. If he types "cd book" then the active directory becomes book and the path name for ch1 becomes simply ch1.

There is a special entry in each directory named ".." (dot dot) that specifies the father of the active directory. For example, if the current directory is "book" then we have:

..   This specifies /usr/rch
../t.e   This specifies /usr/rch/t.e

If we are in the book directory and wish to move to rch's bin directory, we type:

cd ../bin

There is also a "." (dot) entry in each directory that specifies the directory itself.

Both the root directory and the rch directory contain sub-directories called bin. (The name "bin" is historically derived from "binary code.") These bin directories are searched for commands that the user types. For example, suppose rch is in his home directory and types the command "list"; then the directories to be searched for the list command will include both /usr/rch/bin and /bin. The /bin directory contains standard programs such as the "ed" editor. In the case of "list", there is a file named list in /usr/rch/bin, so it will be executed. Each user of Unix can explicitly specify the set of directories to be searched to find commands.

SPECIAL FILES

Unix supports *compatible input/output*, meaning that peripheral devices and files are accessed by the same set of system calls. The devices are considered to be "special" files, and are accessed via entries in directories. By convention, the directory /dev has entries for devices. For example, /dev/mt0 may correspond to a tape drive. To use this tape drive, a user process simply opens /dev/mt0 and reads or writes as if using an ordinary file. Control of access to devices is done by the same method that is used for protecting files.

FILE PROTECTION

Unix provides a simple, but useful scheme of file protection. Each user of Unix has an account name or user ID, say rch, and belongs to a *group*, say osdevel (operating system development group). Each file is considered to belong to an account, and has a string of bits specifying access rights for the file. These bits are represented by a string of letters:

d rwx rwx rwx.

The leftmost rwx triple consists of three bits specifying whether the *owner* is allowed to read (r), write (w) or execute (x) the file. The middle rwx gives the rights of users within the *group* to read, write or execute the file. The rightmost rwx gives the rights of users that are not in the group. The chmod (change mode) command is used by the owner of a file to set its protection flags. For example, the owner can set the access rights to file copy.e by

% chmod 640 copy.e

This uses the octal string 0640, which in binary is:

110 100 000

This gives read/write access to the file's owner, read access to the group and no access to others. If we now type the command "ls -l copy.e", we are given the long (-l) listing (ls) for the file named copy.e. This will print the access rights "-rw-r-----". If you have access to a Unix system, you should try this.

Besides the nine bits for specifying read/write/execute, there are three bits called: (1) set user ID, (2) set group ID and (3) sticky. The third, the so-called sticky bit, informs the system that the file is frequently executed and so a copy of it should be saved in the swap area.

The "set user ID" bit is used primarily to allow critical operating system functions to be executed by user processes. When a file having its set user ID bit on is executed, the user process's account is switched to be that of the owner of the file. If the file owner is the *super user*, an account recognized specially by the nucleus, then the process gains the ability to execute privileged system calls. Seemingly, the original user gains the ability to use these privileged system calls for any purpose, perhaps to the detriment of the system and other users. This is not the case, because the process is now executing a program belonging to the super user; by executing the super user's program, the process carries out the super user's instructions. An example of the use of this technique is in the creation of directories; this is done by a user process that becomes the super user by executing a file with the set user ID bit on. Then the process executes a privileged system call (mknod) to create the directory. The set group ID bit is analogous to the set user ID bit.

Any user can turn on the set user ID bit of one of his files. This file can be executed by other users and can give controlled access to the rest of the first user's files, using whatever access control algorithm is implemented in the file having the set user id bit on.

SYSTEM CALLS TO MANIPULATE FILES

Unix supports communication with files, devices and processes in a simple, consistent manner. Each user process has a number of *channels*, which can be read from or written to. These channels are numbered 0, 1, 2 and so on. There is a maximum number of channels per process; this number is fixed at the time the Unix nucleus is compiled and is typically set to 15.

A process's channels are connected to files and devices by the "open" and "create" system calls. The "pipe" system call connects a channel to another channel to support inter-process communication. The "close" system call is used to disconnect channels.

By Unix convention, channel zero of each user process is its standard input channel and channel one is its standard output channel. A process that is interacting with a terminal has channel zero connected to the user's keyboard and channel one connected to his terminal screen. When a process is created, its standard input and output channels are already connected; the process can immediately begin reading from channel zero and writing to channel one without opening them.

For the rest of this section, we will concentrate on the case of channels attached to disk files, but most of what will be said applies as well to channels that are attached to devices or that are attached via pipes to processes.

When a user process opens a file, the nucleus finds one of the process's idle channels, connects it to the file, and returns the channel number (called a "file descriptor" in Unix terminology). When the process reads or writes the file it does so by invoking the nucleus and passing the nucleus the channel number.

Each read or write is a system call (trap) that invokes the Unix nucleus. The nucleus carries out the requested action and then returns control to the user process. We will now describe Unix system calls used to access files. We will give the calls using the syntax of the C language.

Open. A user process gains access to an existing file by executing:

channelNo = open(fileName, accessMode)

The accessMode specifies reading, writing or updating. (Updating means both reading and writing.) The returned channelNo is the number of one of the process's channels not previously in use. The fileName is the "path name" of the file.

Close. A user process disconnects itself from a file and frees the associated channel by executing:

close(channelNo)

Read. Once a file is open, it can be read as follows:

bytesTransferred = read(channelNo, memoryLoc, byteCount)

The specified number of bytes are to be read from the given channel into the process's virtual memory, starting at the address: memoryLoc. The following values of bytesTransferred are important: (a) if bytesTransferred = byteCount then the requested transfer was done; (b) if bytesTransferred = 0 then the end of the file had been reached and no bytes were transferred; or (c) if bytesTransferred < byteCount then bytes were transferred until end-of-file or end-of-line was encountered. Case (c) commonly happens when a channel is connected to a terminal keyboard; a read from a keyboard transmits at most one line of input.

Write. An open file is written to this way:

write(channelNo, memoryLoc, byteCount)

This is similar to the read system call and transfers the specified bytes.

Seek. Random access to a file is accomplished by explicitly specifying the next location in the file to read or write, as in:

lseek(channelNo, offset, base)

The operation is called "lseek" (long seek) for historical reasons: there used to be an operation called "seek", that allowed offsets only up to 32K bytes. This system call moves the position in the file so it becomes "offset" bytes from the given base. The base can be specified as the beginning of the file, the end of the file, or the current position in the file.

We use the operations just described (open, close, read, write and lseek) to manipulate existing files. We use the following operations (create, link and unlink) for creating new files, giving them alternate names and deleting them.

Create. A new file with null contents is created by:

channelNo = create(fileName, accessMode)

This performs the same action as "open", but creates the file if it does not already exist. (Note: the Unix system call is actually spelled "creat".) If it already exists, it is truncated to zero length. Unlike most other operating systems, Unix does not allocate disk space for a file when it is created. Instead, space is implicitly allocated when a write adds to the end of the file.

Link. An existing file is given an alternate name by linking to it:

link(fileName, alternateFileName)

The contents of the file are unaltered. The file may be opened using any of its names. Both names are given as paths.

Unlink. The "unlink" system call deletes one of a file's alternate names:

unlink(fileName)

When all of the file's names have been deleted and no processes have the file open, it is destroyed. So, a file that was never linked to is destroyed by an unlink.

Some of these operations do not make sense for certain attachments to channels. For example, a channel that is attached to a terminal keyboard can be read from, but cannot be written to. Operations fail for other reasons as well: for example, an open of a non-existent file necessarily fails. When Unix receives a request it cannot carry out, it ignores the request and returns a failure status code. To keep the descriptions simple, we have ignored these status codes.

INTERNAL FORMAT OF FILES

It is typical for an operating system to provide a mechanism for structuring individual files. For example, the operating system may be requested to keep track of the fact that a file represents 80-byte records. Unix's approach is different. It supports only one kind of file: a sequence of bytes. If a user process wants to read 80 bytes at a time from a Unix file, then he does so by the read system call. But the Unix nucleus does not enforce the convention.

There are conventions for internal file formats in Unix, but these are followed by user processes and ignored by the nucleus. For example, an ASCII text file consists of lines, where each line is a sequence of bytes followed by a new-line byte. Conceptually, the ASCII text file consists of a sequence of varying length records; this concept is represented only by the new-line characters in the file and is not enforced by the nucleus.

Another convention is used for files containing executable machine code. Each such file starts with a special word value. This value is unprintable and is unlikely to occur by accident as a file's first bytes; it is used to check if a file that is to be executed actually contains machine code.

Other operating systems distinguish between various internal file organization techniques such as sequential vs. random access vs. indexed sequential. Unix ignores these distinctions and supports only one kind of file, leaving the question of organization to the users. The result is that there is less for the user to learn and less for the operating system to implement.

MOUNTING DISK PACKS

The Unix file system resides on a set of disks. (Actually it could as well reside on other data storage media, such as bubble memory, but we will only refer to disks.) The root directory resides on a particular disk, and the tree of directories begins with directories stored on this device. Access to files on another disk requires a system call to "mount" the other disk; this call effectively replaces a node in the original file system by the tree residing on the disk being mounted. This is accomplished by the system call

mount(nodeName, deviceName, readWriteFlag)

If the nodeName is /usr/rch/test and the deviceName is /dev/disk2, this replaces the test file with the directory tree on disk2. The readWriteFlag specifies whether the device will be allowed to be written on. Once the mount is done, the files on the new disk are accessed as if they were on the original disk. To reverse the effect of mount, we use:

umount(deviceName)

If the deviceName is /dev/disk2, then the file /user/rch/test again becomes accessible. When Unix is initializing itself, it must "mount" any disks that are by default part of its file system.

Each disk pack is initialized to a clean, empty state by a utility program that writes directly to the device, for example, to /dev/disk2. This initialization sets aside space for all files that are to be created on the disk. Unix has a restriction that the space for a file must be allocated on its own disk and cannot use space on another disk.

The link system call (described above) makes an entry in a directory to give an alternate path name for an already existing file. Links are allowed to leaf files but not to directories. Unix has the restriction that all links to a file must be within the same disk.

These restrictions to allocate space strictly within the disk and to keep links within the disk are very helpful for simplifying maintenance. They imply that each disk pack is a self-contained unit that can be mounted, dismounted and checked for consistency without reference to other disk packs. This avoids a considerable amount of housekeeping that would be required to keep track of links and files spanning packs.

We have assumed that each mountable directory sub-tree corresponds to a disk. Although this was originally true, later versions of Unix have allowed single disks to contain several of these mountable sub-trees. In Unix terminology these sub-trees are called "file systems". We have avoided this confusing terminology because the whole of the directory tree together with the algorithms that manipulate files is also called the "file system".

CHAPTER 6 SUMMARY

This chapter has introduced Unix, with an emphasis on its user interface and its file system. Unix is a process-based operating system written in a language called C. It is a small to medium-sized system that is portable and flexible.

The following important terms were introduced.

Interactive users - People who use Unix via terminals.

Commands - Requests that an interactive user can type to a system such as Unix.

Prompt character - The system prints a prompt character on a terminal to signal that it is ready to receive another command.

User processes - Asynchronous tasks that can carry out requests (commands) from interactive users.

Shell - This is a program that interprets users' commands. It executes as a user process. It prints prompt characters and reads commands.

Nucleus - The basic software of Unix that implements user processes and system calls. (In Unix terminology, the Unix nucleus is called the Unix "kernel".)

System calls - These are requests by user processes to the nucleus. For example, a user process can execute a system call to "open" a file.

Accounts - To use Unix, a person needs an account, denoted by a user ID. This user ID is given to Unix when the user logs in at a terminal.

Password - The user must give his secret password when logging in.

Directory hierarchy - The files of Unix are arranged in a tree of directories.

Root directory - The base or root of the directory hierarchy is named "/".

Home directory - Each account has a home directory in the directory tree. By default, the user's files are located in this directory.

Current directory - A user process (and also an interactive user) is considered to be active in a particular directory, which is by default the user's home directory.

Path name - Each file or directory has a "rooted" path, for example /usr/rch/book/ch1, which starts with "/". Alternately, the file or directory can be located relative to the current directory, for example, from the directory /usr/rch we have the path book/ch1.

Search path - The user can specify the directories (search paths) to be searched to find commands.

File protection - Unix uses a string of bits, usually written as d rwx rwx rwx, to specify allowed access to each file. These allow read, write and execute access to the owner, to the group and to others.

Special file - Unix provides compatible I/O between files and peripheral devices by treating peripheral devices as "special files".

Set user ID bit - One of a file's protection bits specifies that a process executing the file is to have its user ID set to be that of the file's owner.

Channel - A user process has a number (typically 15) of potential attachments to files. These "channels" are numbered 0, 1, 2, ... Opening a file attaches it to a channel and closing a file detaches the channel.

Mounting disk packs - A formatted Unix disk pack contains its own directory hierarchy. When a disk pack is placed on a disk drive, it is "mounted" to become a subtree in the system's directory tree.

Following is a list of important Unix commands (which can be typed at a terminal):

who - Types a list of active users.

write - Writes a message on another user's terminal.

ed - Basic line editor, used to create and modify files.

cat - Prints a file on the terminal.

lpr - Prints a file on the line printer.

cec - Concurrent Euclid compiler. This compiles a CE program prog.e, creating an executable version of the program which can be executed by typing prog.out.

cc - C compiler. This compiles a C program prog.c creating an executable version of the program that can be executed by typing a.out.

rm - Removes (deletes) a file.

ls - Lists contents of a directory.

cd - Change directories to be in a new directory.

mkdir - Create a new directory.

rmdir - Remove (delete) a directory.

chmod - Change mode (change protection bits) of a file.

man - Display manual entry for a given command, system call, etc.

Following are the Unix system calls that a user process can execute to manipulate files.

open - Attach a file to a channel.

close - Detach a file from a channel.

read - Read bytes from a channel.

write - Write bytes to a channel. The file grows implicitly when a write extends the file.

seek - Change the read/write position (offset) in a file attached to a channel.

ioctl - This is a catch-all used to "control" a device, for example, to change the speed for a terminal.

create - Create a new, empty file, and attach it to a channel.

link - Give a file an alternate name.

unlink - Delete one of a file's names. A file with no more names is deleted.

CHAPTER 6 BIBLIOGRAPHY

Ritchie and Thompson's article [1974] gives an excellent overview of their design of Unix. Thomas and Yates [1982] give a good introduction to Unix as well as collected references to Unix materials.

Ritchie, D.M. and Thompson, K. The Unix Time-Sharing System. Comm. ACM 17, 7 (July 1974), 365-375. A revised version of this article and related articles appear in the Bell System Journal 56,6 part 2 (July-Aug. 1978).

Thomas, R. and Yates, J. A user guide to the Unix system. Osborne McGraw/Hill, 1982.

CHAPTER 6 EXERCISES

1. If you have access to a Unix system, try out each of the commands listed in the Summary of this chapter.

2. List the system calls of an operating system other than Unix. Compare the system calls for manipulating files with those of Unix.

3. If you have access to a Unix system, read the manual entries for the commands listed in the chapter summary. Note that, for example, "man who" is a command to display the manual entry for "who" on the terminal.

4. Discuss the advantages and disadvantages of Unix's scheme for file protection. Contrast it with simpler schemes, as used on small minicomputer systems, and with more complex schemes such, as Multics' file protection.

5. Unix provides compatible I/O, meaning that file I/O operations also apply to devices. This compatibility is sometimes only partially possible. For example, it is not possible to read from a line printer, or to seek on a terminal. Give a list of such restrictions of compatible I/O applied to "special files".

6. Unix files are created empty, without specifying how large they may grow. What would be gained by knowing the eventual maximum size of a file when it is created. Why did the designers of Unix not require maximum file size at file creation time? Suggest a scheme for allocation of space to Unix-like files as they grow.

7. Give a carefully worded argument explaining how the "set user ID" mechanism allows parts of an operating system to safely execute as user processes.

8. The file containing the encrypted passwords of users can be read by any Unix user. Explain why this does not compromise security.

9. Early versions of Unix did not restrict the directory structure to be a tree. The structure was allowed to be an arbitrary directed graph. Suggest reasons why the designers of Unix decided to restrict it to be a tree.

Chapter 7

UNIX: USER PROCESSES AND THE SHELL

In this chapter we discuss the system calls provided by the Unix nucleus for the creation and deletion of user processes. We show how the command processor (the shell) is implemented using these system calls. The next chapter explains how the nucleus implements system calls.

THE ADDRESS SPACE OF A USER PROCESS

We need to examine Unix's concept of a user process in some detail in order to understand how the shell carries out commands. We start by describing the memory segments addressable by the user process.

Each user process in Unix has its own partition or virtual memory, which we shall call its *address space*. The process can address the bytes in its address space but cannot address other processes' address spaces or the nucleus. A process's address space consists of three parts, called *segments*. These are

1. The *text* (or code) segment, which holds the instructions (program) being executed.

2. The *data* segment, which has an initialized part called ".data" and an uninitialized part that starts out cleared to zeroes.

3. The *stack* segment, which is used to hold the runtime stack of the user process.

The process is not expected to modify its text segment. Given appropriate memory protection hardware to prevent this modification, several processes can share the same text segment. This is done on architectures such as the PDP-11/70.

The initialized part of the data area holds items such as string literals, for example, the string in Concurrent Euclid:

    const s := 'ABC'

or the initialized array in C:

    char s[] = "ABC"

The initialized part of the data segment is stored in a file along with the executable code for the program. This file is called a *load module*. This file does not contain the uninitialized part, but instead just contains a count giving the number of bytes to be allocated and cleared when the program is loaded for execution.

The data segment can be explicitly extended or contracted by the user process by the "break" system call. This system call is used to implement dynamic creation of variables as in the statement:

    coll.New(p)   {Concurrent Euclid}

The stack segment is used primarily to hold data (activation records) required when the user process calls procedures and functions. The stack segment starts out small and implicitly grows as more space is required by the process's activated procedures. The stack segment grows but does not shrink during the execution of a particular program.

An explanation about stacks is in order here. A C program or a Sequential Euclid program uses *one* run-time stack, which corresponds exactly to the stack segment. However, a Concurrent Euclid program with N processes has N stacks. A CE program is run under Unix as a single Unix user process that shares its user processes' time among the CE processes. Its N stacks are implemented as parts of the data area. A CE program running under Unix does not use its partition's stack segment for any of its process's stack.

Besides these three addressable segments Unix uses a fourth segment to hold system data local to the process. This segment is called the *system segment*. This system segment is outside of the user process's address space and the process cannot address or modify it.

Early versions of Unix support memory management by *swapping*, which transfers a user's segments between main memory and the swapping area on disk. Originally, a swap transferred all four segments contiguously out to disk (or back into memory). In later versions of Unix, the text area is shareable among processes, so text segments are swapped separately from the other three segments, which are swapped as a unit.

In paging versions of Unix, only the active pages of a process's partition are kept in memory. (A page is a fixed-size block of information; in typical systems the page size is 1K bytes.) Fortunately, the method of memory management (swapping or paging) is immaterial to the user process, which continues to consider that it has an address space consisting of three segments (text, data and stack).

Swapping works well as long as each process's address space remains relatively small. Since architectures such as the PDP-11 limit addresses to 16-bits, address spaces necessarily remain relatively small. However, with architectures such as the MC68000 and the VAX, addresses are 32-bits, and address spaces can be many megabytes. When address spaces reach this size, paging becomes highly desirable, as it avoids unnecessary transfers of large amounts of information between main memory and disk.

MANIPULATION OF USER PROCESSES

The Unix nucleus supports a set of system calls that allow user processes to be created, destroyed, etc. An existing user process creates a new process by executing:

> processNo = fork()

The creating process receives a number in processNo which uniquely identifies the new process. Here we are using the C language syntax for system calls.

Unix's fork is different from the fork described in Chapter 2 in that in Unix, the fork does not specify where the new process is to begin executing. In Unix, the son continues executing at the same place in the same program as its father. The only difference between the father and the son is that the processNo returned to the son is zero.

Essentially, fork means "clone me". It creates a son user process which has a separate but equal address space, and is identical to the father in all essential aspects except for the value of processNo. In order that these two processes can tell which of them is the father, the fork is used in this manner:

> processNo = fork()   {Zero is returned to the son}
> if processNo not=0 then
> > Carry out father's responsibility
> else
> > Carry out son's responsibility
> end if

In this way, the father and son execute the same text segment but they behave differently. The son has fresh copies of the father's data and stack segments and thus has access to any values which the father had. The son also has the same set of open files as the father.

In many cases, the son process will want to start executing a new program that is stored in a file. To do this, the process executes:

exec(fileName, arg1, arg2, arg3, ...)

This is a system call to the Unix nucleus. It replaces the process's text and data segments with those given by the file. It re-initializes the process's stack segment and places the arguments (arg1, arg2, ...) in this segment. These arguments are null-terminated strings that parameterize the program about to be executed. The father's files remain open for use by both father and son. Exec does not create a new process; it just replaces a process's program and data, allowing the process to execute a new program.

When a father process wants to wait for a son process to terminate, it executes:

processNo = wait(status)

This is a Unix system call and is *not* the same concept as the wait statement of Concurrent Euclid. It blocks the father process until one of its son processes terminates. The identity of the son is returned in processNo. The reason for the termination (an integer) is given by the word pointed to by status.

When a process wishes to terminate its execution (to destroy itself) it executes:

exit(status)

Note that this is a Unix system call and has nothing to do with the Concurrent Euclid exit statement, which terminates a loop. Following the exit, the process ceases to exist; the status is returned to a wait executed by its father. If the father has not yet executed a wait, the son's status is saved until this happens.

The exit system call is used by a process to terminate itself. A different system call, kill, is used to terminate another process:

kill(processNo, signalNo)

The processNo specifies the process to be killed. A process is allowed to kill another process only if it is the super user or if the assassin and victim processes belong to the same account.

Actually, kill is a misnomer and should be called "notify", because in many circumstances, the victim process survives. The signalNo gives an integer which specifies the type of notification. A process that is expecting to receive a notification can execute:

signal(signalNo, procedureAddress)

This is a Unix system call and has nothing to do with the Concurrent Euclid statement "signal". Even worse, it is poorly named; perhaps it should be called "contingency", because it specifies what action is to be taken when the process receives a particular notification (i.e. kill). The action to be taken when notification occurs is one of:

(1) Ignore the notification,

(2) Call a procedure at a specified address, or

(3) Destroy this process when the notification arrives.

By default, action (3) is taken, destroying the process.

The kill and signal system calls are rather crude, but they are useful tools for handling simple synchronization among processes.

IMPLEMENTING THE SHELL

A person using Unix from a terminal types commands to a program called the *shell*. The shell is a command processor which interprets the lines that the user types, and sees that the requested actions are carried out. From the person's point of view, the shell seems *to be* Unix. But from the nucleus's point of view the shell is just a user process, which executes instructions and requests service from the nucleus via system calls.

If the user wants the shell to list the names of files in the current directory, he types the line "ls". If he wants a "long" form of the listing he types the line

> ls -l

The "-l" is the *argument* given to the ls command. It specifies that additional information is to be printed including the last date of modification and the protection bits. Some commands can accept several arguments.

The shell reads this line, separates the command name (ls) from the argument (-l) and searches for a file with the name ls. By default, the search looks in the user's "bin" directory and in the system's "bin" directory. If the command is not found, the shell writes "ls: not found". Since ls is a standard command, we can assume that it will be found.

The shell will have it executed by creating a son process (by forking) and waiting until the son has executed the command. The following is a simplified version of the shell's program:

```
{Shell reads a command and has it executed}
Read command line
Parse command line, isolating commandName and arguments
sonNo = fork()
if sonNo not= 0 then
        {Father shell continues here and waits
            for son to complete}
        processNo = wait(status)
```

else
    {Son shell continues here}
    Search for file named commandName
    if search was successful then
        exec(commandName, arguments)
        {When son shell does this exec, it stops being
           a shell and starts executing the command.
           When the command has completed, the son
           exits and its process is destroyed}
    else
        Print message saying command not found
        exit(failureCode) {Son dies}
    end if
end if
{Shell is ready to read a new command}

Essentially, the father shell calls its son process as a subroutine to execute the command.

The command name and arguments are passed to the son as follows: the father's entire data segment is copied to make the son's data segment. When the son "execs" the command, it replaces its current data, text and stack segments by those for the command. The convention of completely copying the father's data segment by "fork" is straightforward for the nucleus to implement, but it is rather inefficient as most of the father's data is of no use to the son. This copying is avoided in Berkeley's VAX version of Unix by providing a variant of fork that has the son share the father's data and blocks the father until the son does the exec.

INPUT/OUTPUT RE-DIRECTION

When a command such as ls is executed, it assumes that certain of its channels are already initialized. In particular, a command assumes these connections:

Channel 0. This is the *standard input* and is commonly attached to the interactive user's keyboard.

Channel 1. This is the *standard output* and is commonly attached to the interactive user's screen.

In addition the following is often used:

Channel 2. This is the *error output* and is commonly attached to the interactive user's screen.

The error stream is useful for separating error messages from standard output, especially when the standard output is attached to a file.

An interactive user can easily re-direct the standard output, for example:

>   ls > names

This runs the ls command and instead of printing the file list on the user's screen puts the list into a file called "names". If the command accepts input, its standard input can be re-directed, for example:

>   ed filex < script

This runs the Unix editor "ed" to update the file named "filex" using the editor commands in the file named "script".

The shell implements input/output re-direction as follows. When it parses a command line, it recognizes the symbols ">" and "<". The strings following these symbols are taken as the names of output and input files. Before the command is executed, the process's standard output (channel one) and standard input (channel zero) are attached as specified by the command line. When the command executes, it is unaware of this re-direction and in general considers its standard input and output to be a keyboard and screen.

This convenient method of input/output re-direction supported by Unix is particularly powerful in that it allows a program to be easily used either interactively or non-interactively.

BACKGROUND PROCESSING

By default, the shell waits for a command to finish before reading the user's next command. When a command takes more than a few seconds, the wait is tedious and can be avoided. For example,

>   cec test.e > errors &

runs the Concurrent Euclid compiler (cec) on the program test.e and puts the error messages in the file "errors". (Note that compilers under Unix do not in general produce source listings, so the output stream from cec contains only error messages.) The final "&" means that the command is to be done in the background and the shell will immediately accept new commands.

It is very easy for the shell to implement background processing. It simply treats the command as any other command, but does not wait until the son process carrying out the command terminates. The result is that the shell and the son process carrying out the command continue in parallel.

PIPES AND FILTERS

The output stream from a process can be made to be the input stream of another process by connecting the processes via a pipe. For example, the command line

    ls llpr

means to take the output from the directory lister (ls) and to feed it to the printer (lpr, for line printer). This could be done less conveniently by:

    ls > tempfile
    lpr < tempfile

followed by the removal of the file "tempfile". Several commands can be strung together, as in

    ls lpr -2 llpr

In this case, the output of ls is consumed by pr. The pr program paginates its input, using double columns (the "-2" argument to pr means double columns). The output of pr is then fed to lpr which prints it.

We call pr a *filter*, because it copies its input to its output, with certain modifications. Many Unix programs, such as the "sort" routine, are useful filters.

SYSTEM CALLS TO SUPPORT PIPES

We now show in some detail how Unix system calls are used by the shell to support pipes. We will concentrate on the case of a father process (the shell) arranging things so two of his sons communicate via a pipe.

To create a pipe, the father executes the system call:

    pipe(pipeChannels)

The Unix nucleus picks two of the father's unattached channels, connects them to the newly created pipe, and returns the channel numbers to the father. "PipeChannels" is an array indexed from 0 to 1; pipeChannels[1] is set to the channel number to be used for writing to the pipe, and pipeChannels[0] is set to the channel number for reading from the pipe. This seemingly useless arrangement would allow the father to send a message to the pipe and later read it back, but this is not what we want to do.

The reason for creating the pipe is for use by son processes. The father creates two sons, each of which inherits the father's connections to the pipe. By a pre-established convention, one son will produce data and write it to the pipe, while the other son will consume data from the pipe. This arrangement is established in the following program:

```
{Father executes, creating a pipe and two sons}
pipe(pipeChannels)  {Create pipe}
sonA = fork()
if sonA not= 0  {Father and sonA execute test} then
        {Father continues here}
        sonB = fork()
        if sonB not= 0  {Father and sonB execute test} then
            {Father continues here}
            {Father closes pipe channels, waits for sons}
            close(pipeChannels[0])
            close(pipeChannels[1])
            sonIdent = wait(status)    {One son dies}
            sonIdent = wait(status)    {Another son dies}
        else
            {SonB continues here, consuming from pipeChannels[0]}
            close(pipeChannels[1])
            ... read from pipe ...
            exit(status)    {SonB dies}
        end if
else
        {SonA continues here, producing into pipeChannels[1]}
        close(pipeChannels[0])
        ... write to pipe ....
        exit(status)   {SonA dies}
end if
{Father continues here after death of sons}
```

Recall that a fork in Unix copies all of the father's data for the son. Each of the three processes will have its own array called pipeChannels whose value is assigned in the father's call to create the pipe. When the first son is created, its variable sonA is set to zero, while the father's version of this variable is set by fork to be the process identity of the first son.

FILES CONTAINING COMMANDS

Sometimes it is convenient to collect a sequence of commands and have them executed as a unit. For example, suppose the following commands are in a file called testx:

```
cec x.e
x.out
```

This compiles the Concurrent Euclid program x.e putting the compiled program into file x.out, and then executes x.out. Assuming the protection bits for testx specify that it is executable we can type

     testx

and the sequence will be carried out.

We call a file such as testx a *command file* or a *shell file*, and we say it is written in *shell language*. A shell file can contain any commands or operations (such as "<" for input re-direction) that can be entered interactively. Although we will not go into it here, shell language is reasonably general, supporting loops, selection (case and if), parameter substitution and other features. What this means is that under Unix, the notations of an interactive command language and a job control language (JCL) have been merged to form a single convenient notation: shell language.

The shell executes shell files in an elegant manner, which may seem confusing at first. Basically, the shell calls itself recursively to handle a shell file. Consider the case of testx. When the shell reads a line containing the string "testx", it creates a son shell to carry out the testx command. When the son attempts to "exec" testx, it may fail for one of two reasons: (1) the file is not executable due to the setting of its protection bits, or (2) the protection bits allow the file to be executed, but the file does not start with a peculiar unique word indicating that the file contains machine code. In the first case, an error message is printed and the command is ignored. In the second case the son shell assumes that instead of machine code, the file contains commands. So the son shell process simply continues to execute as a shell, reading commands from the file. Once the son shell reaches the end of the shell file, it terminates. One of the beauties of this recursive scheme is that the shell file may in turn contain names of other shell files, which are executed by creating yet another son (a grandson) to handle the new shell file, and so on.

SYSTEM INITIALIZATION

A Unix system is started up, or *bootstrapped*, by loading the nucleus from a specified Unix file. Once the nucleus has initialized itself, it creates a single user process, the "ultimate ancestor" of the family tree of user processes. This first process executes a program called "init". Init opens a file containing a description of each terminal attached to the system. For each such terminal, a son process is created to handle that terminal. The son tries to read from the terminal's keyboard; in the case of a dial-up line, the open blocks the process until the line is actually dialed up. Next the son prints "login:" on the terminal and issues a read to get the user's account name. Once the account name is read, "passwd:" is printed and the password is read. Assuming that the account name and password are legitimate, the process sets its user identification and home directory to those of the person logging in, and executes the shell. Then the shell accepts and carries out commands from the terminal until the interactive session is complete. At completion, the shell exits, destroying its process. The original init process waits for its sons to terminate, and when this happens, it creates a new son process to handle the next interactive session.

The init process's account name is the super user. This allows it to execute privileged system calls, such as "setuid" to change the process's account name. This is called by a son of init before it exec's the shell. Intriguingly, the shell does not execute as the super user and has no special privileges, beyond those implied by the user's account.

Since the shell has no special privileges, it can easily be replaced by another program. For example, if a word processor operator is using Unix, his default "shell" can be changed to be the word processing program. This technique has two advantages. First, it eliminates the need for certain users to need to learn how to communicate with the real shell. Second, it can be used to prevent certain users from having full access to the system, because all commands can be "audited" by the substitute shell.

CHAPTER 7 SUMMARY

This chapter has discussed the Unix system calls that are used to control user processes. Each user process has a distinct address space, which consists of a text (code) segment, a data segment and a stack segment. User processes may share the same (read only) text segment, for example, when both processes are executing the shell.

The system calls for controling user processes are:

fork - Creates a son process with a copy of its father's address space. The return code from a fork call distinguishes the father from the son.

wait - Blocks a father process until one of its son processes terminates. (This is not the same concept as the CE wait statement.)

exit - Executed by a process to terminate itself. (This is not the same as the CE exit statement.)

exec - Causes a process to start executing a new program. New data and text segments are loaded and the stack segment is re-initialized. Arguments are passed in the new stack segment.

kill - Sends a notification (signal) to another process. By default this terminates the other process.

signal - Informs the nucleus of what action to take when a kill or trap occurs.

The shell (command interpreter) is implemented using these system calls. When it receives a command, it creates a son process to carry out the command. The father shell waits for the son to complete, unless an ampersand (&) specifies that the command is to be done in the background. I/O redirection via > and < is implemented by attaching I/O standard channels (0=input, 1=output, 2=error output) before exec'ing a command.

Pipes are implemented using the pipe system call; the shell uses this system call to interconnect the channels of son processes.

A sequence of commands to the shell stored in a file are called a shell file (or shell script). The shell executes a shell file by creating a son to read the script, which in turn creates son processes to carry out individual commands.

Unix initializes itself by creating a single user process, which executes a program called init. Init creates (and re-creates) son processes which control each of the system's interactive terminals.

CHAPTER 7 BIBLIOGRAPHY

The implementation of the shell and "init"' is described by Ritchie and Thompson [1974]. A number of the concepts used in Unix are adapted from Multics [Organick 1972]. McIlroy et al [1978] discuss the Unix style of software development using software tools such as shell language. Bourne [1978] describes the shell language of Version 7 Unix.

Bourne, S.R. The Unix shell. Bell System Technical Journal, 15,6 Part 2 (July-Aug. 1978), pp.1971-1990.

McIlroy, M.D., Pinson, E.N. and Tague, E.G. Unix time-sharing system: foreword. Bell System Technical Journal, 15,6 (July-Aug. 1978), pp.1899-1904.

Organick, E.I. *The Multics system: an examination of its structure.* The MIT Press, Cambridge, Mass., 1972.

Ritchie, D.M. and Thompson, K. The Unix time-sharing system. *Comm. ACM 17*,7 (July 1974), 365-375; revised version in Bell System Journal 56,6 part 2 (July-Aug. 1978).

CHAPTER 7 EXERCISES

1. Unix's fork system call is surprising in that the son process continues executing the same program as the father. Consider a more elaborate version of fork that combines the function of Unix's fork and exec system calls. Why do you think the designer's of Unix's chose the simpler version of fork? What are its advantages and disadvantages. Describe a more elaborate version of fork that allows the shell to be more efficient.

2. The pipe system call attaches one of a process's channels to another of its channels. This seems useless. Explain why the designers of Unix implemented this system call, rather then the more obvious concept of attaching one son's channel to another son's channel.

3. Assuming you have access to Unix or a Unix-like system, write demonstration programs that invoke these system calls: fork, exec, exit, signal and kill.

4. There is one kind of signal (kill) that always terminates a process (assuming the killer process has the same user ID or is the super user). Describe ways of creating "process cancer", i.e., processes that create other processes rapidly and recursively in such a way that it is hard to destroy all these processes. Describe anti-cancer treatment.

5. A pipe is limited in size (typically 4k bytes). A process that fills up a pipe to this maximum size is blocked on its next write to the pipe. Describe the various kinds of deadlocks that can arise when using pipes. Can any of these arise from using the "|" shell notation, or do all of them require explicit use of the pipe system call?

Chapter 8

IMPLEMENTATION OF THE UNIX NUCLEUS

The previous two chapters discussed Unix as seen by interactive users and the Unix nucleus as seen by user processes. (Recall that what we call the Unix nucleus is elsewhere called the Unix kernel.) This chapter looks inside the Unix nucleus to see how it is implemented. We will consider the major data structures manipulated by the nucleus, including i-nodes and user descriptors. The chapter following this one goes into more detail, showing how the Tunis nucleus can be implemented in CE.

LAYOUT OF DATA ON DISKS

Unix's files reside on disks. Each disk is divided into five areas:

Boot block: The Unix system can be started up by loading and executing this block.

Super block: This block specifies the boundaries of the following three areas on the disk. It also contains the head of the free list of blocks available to be allocated to files.

I-node area: This contains descriptors (i-nodes) for each file or directory on the disk. Each i-node is the same size (64 bytes in Version 7 Unix). By convention, the second i-node represents the disk's root directory.

File contents area: This is used to store the contents of files. The free list head in the super block keeps track of unallocated space in the file contents area.

Swap area: This holds user processes when they have been swapped out of main memory. In Version 7 of Unix, swaps are to a single disk, so only one disk needs to contain a non-null swap area.

Some Unix installations arrange things so one physical disk contains more than one logical disk. With this arrangement, a single physical disk contains more than one of the layouts just described.

THE FLAT FILE SYSTEM VS. THE TREE FILE SYSTEM

The implementation of Unix's file system is divided into two distinct layers. We will call the inner layer the *flat file system* (or *i-node file system*) because it has no directories; flat files on a particular disk have as their names: 1, 2, 3, etc. Each number tells which i-node defines the file. Files in the flat file system are created empty and grow when they are extended by "writes", consuming space from the file contents area of the disk.

The outer layer is the *tree file system* (or *directory file system*). This layer uses the flat file system to implement Unix's hierarchic file directory. It uses each flat file to represent: (1) a directory in Unix's file hierarchy or (2) a user's ordinary disk file.

FORMAT OF DIRECTORIES

A directory is a flat file and consists of 16-byte entries. Each entry consists of a 2-byte i-node number and a 14-byte file name. For example, here is a directory:

i-node number	File name
152	.
18	..
216	myfile
4	oldfile

Each entry is used to map a file name to its corresponding i-node. In this example, oldfile is represented by flat file number 4 on this disk. The dot-dot (..) entry locates the i-node for the father directory of this directory. The dot (.) entry locates the i-node for the present directory, i.e., it is a self reference.

FORMAT OF I-NODES

The contents of a file are kept separate from its control information, which means that a file's contents and its i-node are not adjacent on the disk.

Each i-node consists of fields giving information such as the file's owner, its size and its date of creation. Here are the fields of an i-node:

User number: uniquely determines the owner's account name (user ID).

Group number: used for file protection.

Protection rights: the string of protection bits.

Times: when file was last read and last updated and when i-node was last updated.

File code: specifies if the i-node represents a directory, an ordinary user file or a special file. (A special file is typically a peripheral device.)

Size: length of file in bytes.

Block list: locates contents of file.

Link count: gives number of directories referencing this i-node.

All of these fields except for the file code and the link count can be considered to belong to the flat file system. It is the tree file system's responsibility to decide whether an i-node represents a directory and to record how many directory entries reference (are linked to) a particular file.

If the file code indicates a special file, then the i-node does not represent a disk file. Instead, the i-node is being used for a special purpose, typically for accessing a peripheral device. For example, each terminal on a Unix system is accessed via an i-node.

BLOCK LISTS

The block list of an i-node consists of pointers to blocks in the file contents area of the disk. In Version 7 of Unix, there are thirteen of these pointers (Version 6 uses a different arrangement). This table shows an i-node's block list.

Disk Accesses	Byte Range	Block Pointer	
		1	
		2	Direct
1	To 5120	3	
		.	
		.	
		.	
		10	
2	To 70,656	11	Indirect
3	To 8,459,264	12	Double indirect
4	To 1,082,201,087	13	Triple indirect

I-NODE BLOCK LIST

The first ten pointers directly locate file content blocks on the disk. Each block is 512 bytes, and these pointers locate the first 5120 bytes of the file. The eleventh pointer is used when the file size exceeds 5120 bytes. It locates a block which in turn contains pointers to blocks containing the file's data. Thus, bytes of the file that are located via the eleventh pointer require an extra disk reference. Beyond byte number 70,656 we use the twelfth pointer, which has double indirection; it locates a block containing pointers to blocks of pointers to blocks of data. Beyond byte number 8,459,264 we go to pointer number thirteen and triple indirection, which supports an absolute maximum file size of 1,082,201,087 bytes.

The amount of indirection increases with the size of the file; for very large files, in principle we require up to four disk accesses instead of one for small files. Fortunately, there is a software cache that holds recently referenced blocks, so in many cases the required intermediate pointer blocks are already in memory. (Note: Berkeley Unix has used a similar arrangement but its block size is 1024 bytes.)

Before discussing the file system's data structures in more detail, we need to examine the descriptors used to represent user processes.

DESCRIPTORS FOR USER PROCESSES

The Unix nucleus manages certain information about each user process. This "per process" information is kept in a pair of data structures:

(1) Process descriptor. This is permanently resident in memory.

(2) User descriptor. This contains information that is needed only when the user process is in main memory. This descriptor occupies the system segment, one of a process's four segments, which is swapped between memory and disk. Recall that the other three segments (text, code and stack) constitute the process's address space.

The size of the process descriptor is minimized in Unix so that inactive processes, such as those waiting for log-ins, occupy little memory.

Each user process has a number (its *process ID*) which is a pointer into the table of process descriptors. When a fork is executed, the father process is given the number of its son. In the kill system call, the victim is specified by giving its process number. From each process descriptor we can locate its user descriptor, which will be either in memory or swapped out to disk.

There are fields in the user descriptor for the process's user ID, current directory, general registers and status of files. We will now trace the path from the file status fields in the user descriptor to i-nodes on disk.

LINKAGE FROM USER PROCESSES TO DISK FILES

Each user process has a number of channels which can be attached via i-nodes to disk files, special files or pipes. Each user descriptor contains a *channel table* (also called the "per user open file table"). When the process executes an open or create system call, an entry in this table is filled in and the number of this entry is returned as the channel number. This entry contains a pointer to an entry in another data structure, called the *open file table*. In turn, the open file table entry has a pointer to an entry in the *memory resident i-node table*. This chain of pointers is illustrated here:

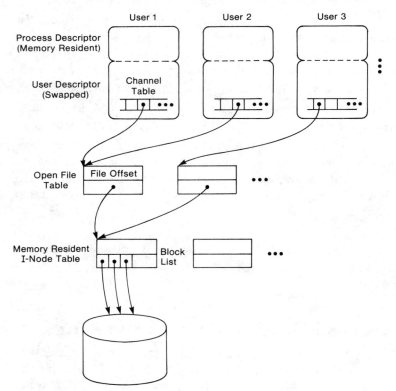

We will trace this chain from bottom to top starting with i-nodes. Consider the case when a disk file is opened. An entry is allocated in the memory resident i-node table and the disk resident i-node is read into this entry. The memory resident version of the i-node will be updated to reflect any changes in the disk file and does not need to be written back onto disk until the file is no longer active. When a user process opens a file that is already active, it shares the existing entry in the memory resident i-node table.

When the file is opened, an entry is also allocated in the open file table. In one field of this entry, a pointer is created to locate the corresponding memory resident i-node. In another field is the entry's *file offset* (or file pointer); it is set to zero. It locates the next byte of the file to be read or written. Each read, write or seek accesses this open file table entry and updates its offset field.

The flat file system does only random access I/O; it is passed this offset field with each read or write. A seek on a file changes this offset field, without accessing the i-node.

When a user process forks, its son process inherits the father's open files. This is implemented by making a copy of the father's channel table for the son. As a result, the father and son share the same open file table entries. Whenever either process does a read, write or seek on one of these files, the shared offset field in the open file table is updated. This rather surprising arrangement has been set up for a purpose. It makes it possible for the father to begin processing a file, to give it to a son for further processing, and then for the father to continue with the remainder of the file. Unfortunately, if either process does internal read-ahead buffering of the file, this arrangement does not work.

The items along the path from the channel table to the file are de-allocated when no longer needed. There are usage counts for the various table entries that record the number of activities that require the entry. When a usage count goes to zero, the table entry is freed.

An open file table entry is de-allocated when a close sets its usage count to zero. Similarly, a close may reduce the activity of a memory resident i-node to zero. As a result, the updated i-node is written to disk and the entry in the memory resident i-node table is freed. When a file is removed (unlinked), the link count of its i-node may go to zero. If so, the disk resident i-node is freed; it is written to disk with a zero link count.

Things get confusing when an open file is unlinked. In this case the i-node's link count can go to zero, but the file's i-node is still being used. As long as the file stays open, the memory resident i-node's usage (activity) count remains positive. Therefore, the file remains in the flat file system, in spite of the fact that the file is no longer part of the tree file system (because it has no path name). This anomaly is resolved when the file is closed and is removed from the flat file system by writing its defunct i-node to disk.

We have thus far glossed over the buffering that makes the file system efficient. There is a software cache of recently read/written disk blocks. If a requested block is already in memory, it is used, thereby avoiding physical input/output. Analogously, there is a cache of unused i-nodes, so it is not usually necessary to access the disk when creating a new file. The free list of available blocks for each disk consists of a linked list whose elements are blocks, each containing 50 pointers to free blocks. The first such block for each disk is kept in memory. As a result, a new block can usually be assigned to a file without accessing the disk.

LINKAGE FROM USER PROCESSES TO SPECIAL FILES

The compatible input/output of Unix provides access to peripheral devices as if they were files. If the file is special, the block list is meaningless, so the block list fields are used for other purposes. In particular, new fields record the device's ID, which consists of two parts:

Major device number: specifies the class of device, e.g., the controller managing a set of tape drives.

Minor device number: specifies the device within the class, e.g., the particular tape drive.

A special file is opened much like an ordinary file. Its disk resident i-node is brought into memory, an open file table entry is pointed to it, and a channel table entry is pointed to the open file table entry. When the user process executes an operation such as read or write on the special file, the nucleus notices from the memory-resident i-node that the file is special. So, the nucleus extracts from the i-node the device ID and uses the major device number as an index into a data structure called the *configuration table*. This table locates a procedure within the nucleus that carries out the desired operation for the particular special file. This procedure is called with the minor device number as a parameter; it carries out the requested operation and returns.

This method of handling special files is powerful and flexible. It allows new peripheral devices to be attached to Unix system with little difficulty. The new device driver code is written, entries for its procedures are added to the configuration table, and the Unix nucleus is re-compiled. Also, an i-node must be created with the appropriate device ID; this is done by the mknod system call.

In addition to the obvious use of special files to represent peripheral devices, they serve other important functions. For example, a user process can use a special file to by-pass the file system and access each disk as a sequence of blocks. Of course, the protection bits prohibit all but a few users from this dangerous access. But this direct access is important for several purposes. It allows a Unix utility program executing as a user process to layout new disk packs so they can be used to contain Unix files. It also allows a user process to check a Unix disk pack to see if its file system data structures are in a consistent, non-corrupt state.

Special files are also used to implement pseudo-devices. For example, the file /dev/mem by convention represents main memory. A user process with access to this file can read or write any byte in the computer's physical memory. Obviously this ability is not to be used casually, and writing is rarely done. Reading provides a sneak path to the nucleus's data structures. This path is used by some programs to gather system statistics. For example, the ps (process status) program runs as a user process and reads the nucleus's process descriptors to print the status of user processes. Of course, a program such as ps depends intimately on details inside the nucleus and will fail if the nucleus is significantly modified. The ps program would have to be re-written if the Unix nucleus were replaced, for example, by Tunis.

LINKAGE FROM USER PROCESSES TO MOUNTED DISK PACKS

When Unix is started up, it has access to only one disk. It assumes a particular i-node on this disk represents the file system's root directory. Further disks are made part of the directory tree by mounting each new disk on a leaf of the existing tree. The mount system call marks the leaf's memory-resident i-node as "mounted". When the nucleus is following the path of a file name (e.g. to open the file), it may encounter an i-node flagged as mounted. When this occurs, the nucleus searches the *mount table* to find the device ID of the disk whose directory tree is mounted on the i-node. (The mount table contains entries that map mounted i-nodes to corresponding disk device IDs.) The root directory of the new disk effectively replaces the mounted i-node. The open system call may encounter several mounted i-nodes in the course of following a path through the directory tree. Once the desired file is located, its i-node is brought into memory. Subsequent high frequency operations, such as reading, directly reference the memory resident i-node without concern for the path by which it was located.

FILE SYSTEM CONSISTENCY

Ideally the hardware and software for a Unix installation never fail, and there is no difficulty with lost or corrupted files. Unfortunately, life is not this simple and problems do occur. For example, a dip in the electric power voltage may cause the CPU and/or its bus hardware to fail. Such a failure is particularly insidious in Unix because key parts of the structure of the file system are cached in memory. When a sudden failure occurs, these parts are not written out to disk, leaving block lists and directories in an inconsistent state on the disk.

There is a system call (sync) that forces memory resident disk information to be written out to disk. This operation is invoked about every 15 seconds, so that the disk is never far out of date. Sync is also called before shutting down the system to guarantee that the disks are left in a consistent state.

An unexpected hardware failure does not generally permit a sync to be completed. For some failures, such as traps inside the nucleus, it is possible to write cached information to disk. Unfortunately, this is not always a good strategy, because the software failure may have corrupted the memory resident data, and writing this to disk only compounds the problem.

When a Unix system starts up, or when disks are considered suspicious, programs are run to check the disks' consistency. We will discuss two of these programs, called i-check and d-check. The first checks the consistency of the flat (i-node) file system and the second checks the tree (directory) file system. Since each disk is self contained (there are no cross-disk links or multiple disk files), these programs can check one disk at a time.

Among other things, i-check tests to see: *Is each block on exactly one list?* In more detail: Does the free list together with the block lists of allocated i-nodes contain exactly the blocks in the file contents area, with no repeats? This check can be carried out efficiently using a bit vector initialized to zero, where each bit represents a disk block. The free list and the i-node block lists are followed. Each time a block is encountered, its bit is inspected. If the bit is off, it is turned on. If the bit is already on, there is a serious problem because the block is on multiple lists. A block that is both on the free list and allocated to a file can be removed from the free list with no harm done. A block that is allocated to more than one i-node file represents a serious corruption of the file system; this problem has no general, fool-proof cure. Once all lists have been followed, the bit vector is scanned for remaining zeros. Each zero represents a block that is not on a list. This problem is not so serious, because it can be cleared up by adding the block to the free list, on the presumption that it was a free block lost when the cache of free blocks was lost.

The d-check program verifies that the directory structure is legitimate. Among other checks it asks the questions:

Do the directories form a tree?

Does the link count of each file equal the number of directory links to it?

These questions can be answered by a single traversal of the directory tree followed by sequentially reading the i-nodes. We use a zero-initialized vector of counters, one counter for each i-node.

When the traversal encounters a reference to a file named dot dot (..), it verifies that this is a link to the father (or a self link in the case of the root). When dot (.) is encountered, this is checked to verify that it is a self reference. When the directory entry references a son directory, the corresponding counter is inspected. If it is zero, things are as they should be. If it is not zero, there is a problem, because we have previously found a father for this son directory. This inconsistency can be cleared up by deleting all but one father, but this has the risk of losing the legitimate path to the son directory. For all directory entries (including dot and dot dot) we increment the corresponding counter in our vector.

Once the tree traversal is complete, we sequentially read the i-nodes from disk to verify that their link counts are equal to the corresponding counters in our vector. A match indicates consistency. If the vector's counter is non-zero and does not agree with the link count, we restore consistency by setting the link count to vector's counter, with little danger of losing valuable information. But if the vector's counter is zero and the link count is non-zero, then apparently the i-node has gotten detached from the directory tree. We can restore consistency by freeing the i-node and its disk blocks, but this has the danger of destroying valuable files. We will not go into strategies for recovering these orphan files and directories.

Anytime a file is lost or corrupted, the reliability of the system is degraded. Even though Unix provides software tools for patching up tangled data structures on disk, these require a sophisticated system programmer to clean up the mess without losing valuable files.

Much of the danger of file corruption could be avoided by doing away with software caches and immediately updating the disks whenever a change occurs. Unfortunately, the loss of performance implied by this strategy is prohibitive. A more sophisticated strategy is to allow caching but to impose an ordering on the writing of blocks to disk. This ordering should imply that files are never lost or corrupted. Due to caching, after a crash, a particular file may not be completely up to date, but should be in a state such as the file had a few seconds before the failure. (Even this degree of reliability is not sufficient for certain applications such as data bases; to improve reliability in these applications, Unix could provide new system calls to specify the order of file updates.) Newer versions of Unix such as

Berkeley Unix, attempt to minimize file damage by taking care in the order of disk writes.

CONCURRENCY IN THE UNIX NUCLEUS

The Unix nucleus must handle two sources of concurrency: the peripheral devices that run in parallel with the CPU and the user processes that share the system's resources. The devices interrupt the CPU at unpredictable times. The user processes are logically concurrent, meaning that they appear to execute in parallel. Since there is only one CPU, this appearance is supported by time slicing the CPU to share it among the user processes.

In Chapter 2 we suggested two extremes of handling concurrency in operating systems. In the first, a monolithic monitor is used that disables all interrupts while the operating system is executing. Interrupts are enabled only when user processes are executing. The result is that concurrency is easy to manage, because the operating system is only entered at well defined times, namely, when a user process is executing. Unfortunately, this arrangement is not workable for a system of the size of Unix because interrupts would be disabled too long, causing poor response and poor resource utilization.

The second extreme is to have a small synchronization kernel, as is used by CE, and to disable interrupts only in this kernel. Essentially, this is the approach taken by the Tunis kernel, described in the following chapter. The Tunis approach uses CE processes and the DoIO operation to handle devices. Tunis uses CE processes called envelopes to track the execution of user processes and to carry out their requests.

The situation of the Unix nucleus falls somewhere between these extremes. There are no true processes in its nucleus (although they are called "processes" in Unix terminology). Rather, there are second-class processes, which are actually coroutines. When a user process executes a trap, it enters the nucleus and becomes a coroutine. What distinguishes *coroutines* from true processes is the fact that a coroutine never gives up the CPU except by an explicit action on its part. So coroutines do *not* execute logically in parallel. Since Unix supports only one CPU, an executing coroutine knows that no other activity can interfere with it until it gives up the CPU. Unfortunately, this is not really true, because an interrupt can pre-empt the CPU. For the minute, we will ignore the unpleasant reality of interrupts.

There are operations that coroutines execute to put themselves to sleep and to wake up other coroutines. These are:

sleep(E): Wait for event E. The coroutine remains inactive until the event is signaled (by a wakeup) by another coroutine. Following the signal, the highest priority ready coroutine begins to execute.

wakeup(E): All coroutines waiting for event E are moved to the list of ready processes. If no processes are waiting for E, the wakeup has no effect.

The explicit transfer of control by coroutines, as is done by sleep, is sometimes called a *resume*.

Each event E is simply an integer value, for example, we could have sleep(14). When a coroutine executes sleep(E), an entry is made in the nucleus's *event list*. This entry records the identity of the coroutine and the value of E. When a coroutine executes wakeup(E), the event list is searched for all occurrences of value E; each such entry is removed and the corresponding coroutines are transferred to the priority-ordered list of ready coroutines. When a coroutine resumes execution it does not in general know if the status it was waiting for holds. The problem is that the wakeup may activate several coroutines, and besides, higher priority processes may have previously executed. A common pattern of waiting is the following:

```
loop
    exit when status is as required
    sleep(E)
end loop
```

In CE, this looping pattern is not necessary, because signaling a condition immediately wakes up exactly one of the waiting processes; so the waked up process is assured that the awaited status holds. While Unix's sleep and wakeup operations allow coroutines to interact, they do not provide a mechanism for handling interrupts, so we now consider this problem.

HANDLING INTERRUPTS

In the Unix nucleus, each source of interrupt has an *interrupt handling procedure* that is called when the interrupt occurs. By convention, when this procedure is activated, further interrupts from the same source are disabled. This allows the procedure to handle the interrupt and update associated data structures without fear of interference from another related interrupt. Essentially, this disabling provides the procedure with a critical section.

Since interrupts directly or indirectly serve the needs of user processes, the associated data structures are shared with user processes' coroutines. A coroutine that accesses these data structures must be wary of race conditions due to unpredictable invocation of interrupt handling procedures. This problem is solved by providing operations so coroutines can explicitly disable and enable selected interrupts. Essentially these operations implement critical sections to guard against unwanted concurrency.

The interaction of coroutines and interrupt handling procedures can become complex. For example, consider the following situation:

A coroutine representing a user process needs a character from a device. It inspects the data structure that holds such characters and concludes that a character is not available. Then it sleeps until it is waked up by an interrupt handling procedure.

The question is, exactly when should the coroutine disable interrupts. If the data structure is non-trivial, its inspection requires a critical section and hence interrupt disabling. If the interrupts are enabled after the inspection and before the sleep, then this allows the interrupt handling procedure to receive the character and do the wakeup before the sleep occurs. This would be unfortunate because the coroutine would then go to sleep, having missed the character.

Apparently, the coroutine must execute the sleep with the interrupt disabled, with the assumption that the sleep enables the interrupt. The waiting loop must be bracketed by operations that disable and enable the interrupt. We end up with a structure that looks a little like mutexbegin/end as discussed in Chapter 2. However, now we have a sleep in the middle that is a bit like a monitor wait in that it allows other activities to enter critical sections. This rather complex arrangement would be easier to understand and get correct by incorporating it in to a programming language. It has been a goal of languages like CE to formalize these arrangements into reliable programming language constructs.

The strategy of carefully turning on and off interrupts has been used in many programs such as operating systems that manage peripheral devices. Certainly the strategy can be made to work, and it does work in the Unix nucleus. However, it requires a great deal of programmer discipline and sophistication. It is an error-prone strategy that leads to programs that are difficult to understand and maintain. The trouble is that it is not easy to see from the structure of the program what is happening.

By comparison, when using language constructs like those provided by CE, the program structure makes obvious the units of concurrency (CE processes) and the critical sections (monitors). Besides greatly simplifying the structure of the nucleus, this organization has other advantages. One is that concurrency and critical sections are expressed in a machine independent manner. As a result, concurrent system software can be easily tested in a machine independent manner using a simulation kernel. This machine independence implies that the software will be more easily portable to other computer architectures. Another advantage is that with logically concurrent processes (instead of coroutines) it is easy to support multiple CPUs. The reason is that process-based software already presumes that each process has a virtual CPU. Adding more physical CPUs just increases the speed of processes without affecting the correctness of the system. By contrast, a program such as the Unix nucleus that explicitly disables and enables interrupts is difficult to adapt to a multiple-CPU environment.

CHAPTER 8 SUMMARY

This chapter has concentrated on the major data structures that are used in implementing the Unix nucleus. These structures are:

Layout of disk data. Each (logical) disk has five areas: boot block, super block, i-node area, file contents area and swap area.

Free list. This locates the unallocated blocks in the file contents area. Its header lies in the super block.

I-nodes. These are fixed length descriptors for flat files. The fields in an i-node include: user and group numbers, protection bits, times of file activity, file code (ordinary, directory or special), file size, block list and link count.

Block list. For each i-node, this locates the blocks containing the file's data.

Directories. Each directory is a "flat file". A directory consists of 16-byte entries that map file names to i-node numbers.

Descriptors for user processes. Each user process has a permanently memory resident "process descriptor" and a swappable "user descriptor".

Channel table. This is located in the process's user descriptor. Each entry represents a potential attachment to a file. An active entry points to an open file table entry.

Open file table. An "open" system call creates an entry in this table. The entry contains the current "offset" into the file and a pointer to the corresponding entry of the memory resident i-node table.

Memory resident i-node table. Each active file has its disk-resident i-node copied into an entry in this table. This entry is written back onto disk when the file is no longer active.

Configuration table. Locates procedures (open, read, write, close and ioctl) that support I/O for each special file. The major device number, stored in the special file's i-node, is used as an index into this table.

Mount table. This maps each "mounted i-node" to a corresponding device (usually a disk). The directory tree on the device effectively replaces the mounted i-node.

Event table. Within the Unix nucleus, concurrency is handled by coroutines. These coroutines can wait (sleep) by waiting on "events" that are recorded as entries in this table.

CHAPTER 8 BIBLIOGRAPHY

The data structures of the Unix nucleus are described in articles by Ritchie and Thompson [1974,1978]. Analogous data structures for other operating systems are discussed in books such as those by Shaw [1974] and Lister [1975].

Lister, A.M. *Fundamentals of operating systems.* MacMillan Press, 1975.

Ritchie, D.M. and Thompson, K. The Unix time-sharing system. *Comm. ACM 17,*7 (July 1974), 365-375; revised version in Bell System Technical Journal 56,6 part 2 (July-Aug. 1978).

Shaw, A.C. *The logical design of operating systems.* Prentice-Hall, 1974.

Thompson, K. Unix implementation. *Bell System Technical Journal 56,*6 part 2 (July-Aug. 1978). pp.1931-1946.

CHAPTER 8 EXERCISES

1. Many operating system try to allocate files contiguously on disk to cut down on time for disk head movement. Invent algorithms and data structures that attempt to keep Unix files somewhat contiguous.

2. Describe difficulties that would be encountered if i-check and d-check were run on a disk that is actively being updated. Give guidelines for designing a disk consistency algorithm that allows concurrent updating. (This is not trivial!)

3. Describe advantages and disadvantages of (1) multiple-disk files and (2) cross-disk file links.

4. It has been proposed that the first part of a Unix file should be adjacent to its i-node. What are the advantages of this scheme?

5. The link count in an i-node is redundant because it simply records the existing number of links to the file. Since this count is redundant, why

does it appear in the i-node?

6. What are the reasons that all openings of a particular file share the same memory resident i-node, rather than making a new copy?

7. Attempt to design a method of adapting Unix's scheme of coroutines to support a multiple-CPU, shared memory version of Unix. What are the main problems with making this adaption?

Chapter 9

TUNIS:
A UNIX-COMPATIBLE
NUCLEUS

In the preceding chapters we presented Unix without explaining how its nucleus is implemented. We discussed Unix's command language, its hierarchic file system, its method of user process manipulation and various system calls implemented by its nucleus. In this chapter we present the internal structure of a Unix-compatible nucleus: Tunis.

The Unix nucleus (or as it is called elsewhere, the Unix kernel) was originally written in assembly language and then re-implemented in the C language. It was first used widely on the PDP-11. The fact that it is written in C has allowed it to be ported to new architectures including the Interdata 832, the VAX and the Z8000. The function of the Unix nucleus is defined primarily by the system calls it supports for user processes and by the disk resident data structures it manipulates. The Tunis nucleus supports the same system calls and uses the same disk resident data structures so we say it is Unix-compatible at those two interfaces. However, Tunis is written in Concurrent Euclid rather than C, and its internal structure is completely different from that of the original Unix nucleus.

WHY TUNIS?

Tunis (Toronto UNIversity System) evolved out of a project in a graduate course on operating systems at the University of Toronto. The first course project, in Fall 1979, produced a design and a preliminary implementation written in the Toronto Euclid language. In Fall 1980, a more ambitious version of Tunis was designed, this time with the goal of being compatible with Unix. Much of the system was implemented, and inevitably it was discovered that some internal interfaces were ill-conceived. Meantime, this effort lead to the design of the Concurrent Euclid language, which is essentially the full Euclid language as trimmed down to decrease complexity and ease implementation, and beefed up to handle real systems programming problems such as those encountered in the Tunis work. Patrick Cardozo and Mark Mendell designed and implemented major parts of Tunis as a part of their M.Sc. research. Tunis has continued to provide the focus of graduate research; for example, in Fall 1981, a course project investigated running Tunis on the VAX using paging. This type of work continues.

One of the purposes of the Tunis project has been to develop high level software structuring techniques. These techniques include: tightly disciplined programming constructs as in Pascal-like strong type checking; concurrency as in CE's processes and monitors; and enforced modularity as provided by CE's modules.

TENETS OF SOFTWARE ENGINEERING

Designing a program such as the Tunis nucleus is not easy. The traditional challenges to the software engineer are present, including these questions:

How do we divide a large program so that it can be developed in parallel by more than one person? (The development question)

How can we make the program efficient enough for its purposes? (The performance question)

How can we keep the program understandable so that it is easy to maintain? (The maintenance question)

How can we make the program portable so that it can be used on a new computer architecture? (The portability question)

These issues are important to many large programs, but are particularly challenging in the case of an operating system nucleus because performance is critical and because so much concurrency is present.

The traditional tools used to design software include the following. We attempt to divide the program into "good" modules, where a module is considered good when its specification can be simply stated, independent of its implementation. Perhaps the best way to discover good modularity is by *functional decomposition,* which means dividing the program into parts based on the various functions the program is to perform. For example, the part of Tunis that implements file updating should be in a separate module from the part that implements process forking.

From module to module, we strive for *limited visibility*, so each module knows as little as possible about other modules. In CE, this means we try to shorten import/export lists as well as limiting the complexity of imported/exported items. This is called *information hiding* and means that the interfaces between the different parts of a program are kept simple. Sometimes we are faced with a trade-off between narrower interfaces and better performance. This trade-off occurs when access to more information allows more effective scheduling/optimization. This trade-off decision is quite difficult because in a system like Tunis, both modularity and performance are essential.

Within an implementation we strive for *textual isolation* of design decisions. For example, in Tunis the scheduling algorithm that allocates CPU time among processes is located completely within the kernel. If we make a new decision to schedule the processes by a different algorithm, we know that the kernel program is the only text we need to modify.

We should design a program such as an operating system nucleus as *layers*, much as Unix consists of layers. In Unix, the interactive user deals only with the outermost software layer; the interface to this layer is defined by shell commands, editor commands and other terminal oriented interactions. This outermost software layer consists of user processes. These processes access the next layer of Unix, which is the nucleus. The Tunis nucleus is implemented as a number of layers, which are implemented as CE modules.

A *layered* system, such as the T.H.E. operating system or the Tunis nucleus, can be thought of as *levels of abstraction*. This means that each new layer, from the inside out, implements new operations that can be used by successive layers. For example, the nucleus provides operations to create, read, write and destroy files; given the nucleus layers, we can reason in terms of the "file" abstraction. Without the nucleus we would have to reason in terms of lower level abstractions such as blocks of bytes on disk packs. When we design each new layer, we can treat the operations provided by the lower levels as instructions of an *abstract machine*. The new level is not concerned with *how* this abstract machine is implemented, but rather only with *what* it does.

The Tunis nucleus must manage the concurrency of asynchronous devices and user processes. The software structuring technique used inside the Tunis nucleus to control this concurrency is the CE process. These CE processes inside the nucleus are called *system processes*. Since Tunis is written in CE, the obvious way to handle concurrency is to use one process for each unit of asynchronism. For example, each terminal keyboard operates asynchronously, so each keyboard is controlled by a CE process.

These basic concepts of software engineering, including information hiding, levels of abstraction, and handling concurrency by processes, were used to guide the design of Tunis. This can be seen in the structure of Tunis; it consists of layers of modules, each optionally containing processes, all based on a kernel that supports CE concurrency features.

THE LAYER STRUCTURE OF TUNIS

The Tunis nucleus consists of a CE program supported by a kernel written mostly in assembly language. This CE program consists of layers that are CE modules. It is compiled in parts, module by module. The compiled modules are linked with the kernel to make an executable nucleus.

We will present a somewhat idealized version of Tunis's layer structure. Its major layers starting from the outside are:

User manager. Interprets systems calls from user processes and makes calls to lower levels to carry out the user's requests.

File manager. Implements the file system, using devices supported by a lower level.

Memory manager. Shares the available physical memory among user processes to support their address spaces (virtual memories) and uses disk support provided by the next lower level to store inactive user processes' address spaces.

Device manager. Contains device drivers for each peripheral device attached to the system.

Utility layer. Supports common facilities needed by the above layers, in particular:

> Clock manager: allows delaying and reading the time.
> Physical Memory manager: supports transfers of
>   strings of bytes within physical memory.
> Panic manager: shuts down system following disasters.

Kernel. Handles interrupts and supports concurrency as implied by CE language constructs.

We will discuss these layers starting with the bottom.

The kernel is relatively small (between 2K and 4K bytes) and is inherently machine dependent. Its speed is critical to the efficient execution of the system. For these reasons it is written mostly in assembly language. The techniques for constructing a kernel are given in the last chapter of this book. We can think of the kernel as extending the existing hardware to support the concurrency constructs of CE. It is as if we are building a better version of the hardware with more convenient handling of process/process and process/device interactions. Once we have implemented the kernel, we can ignore it and reason more conveniently in terms of CE's concurrency constructs.

The utility layer of Tunis can be thought of as a subroutine package that supports operations needed by upper layers. Its Clock manager is relatively simple, allowing system processes to read or re-set the clock and to wait a given number of seconds. Most entries to the Panic manager simply record the nature of the disaster on the system console and halt the system. The Tunis nucleus executes in a permanently memory resident address space. Depending on the underlying hardware, addresses in this space may or may not be the same as physical addresses. The Physical Memory manager implements transfers between this system address space and physical memory, and from physical memory to physical memory.

THE MAJOR LAYERS

Ignoring the kernel and the utility layer, Tunis has the following structure.

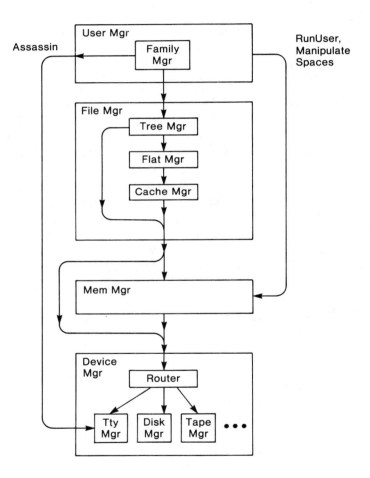

Major Layers of Tunis

As can be seen in the diagram, the major layers have internal structure. Inside the Device manager layer there is a Router that accepts all I/O requests and passes them to the appropriate submanager, such as the Disk manager. Within these submanagers I/O is performed via a Begin/Wait/EndIO as described in Chapter 4.

Inside the Memory manager is a Lock manager that forces all or part of a user's address space into memory. This is necessary, for example, when a DMA (direct memory access) device does I/O directly to/from a user process.

Inside the File manager are three submanagers. The lowest one is the Cache manager, which avoids disk access when a recently accessed disk block is still available in its pool of buffers. Next is the Flat File manager, which uses i-nodes to implement a non-hierarchic file system. Next is the Tree File manager, which uses flat files to implement Unix's hierarchic file system. The Tree File manager determines if a request is to manipulate an ordinary file or a special file. If the file is ordinary, the request is passed to the Flat File manager. If the file is special, then its request is shunted around the Flat File manager. It is passed to the Memory manager, which locks down user address space as needed and subsequently passes the request on to the Device manager.

The highest layer is the User manager. It contains a number of envelope processes. Each envelope acts as a "guardian angel" for a user process. The envelope process waits for a user process to be "born" via a fork. When this happens, the envelope takes control and ushers the user process through its lifetime, handling any system calls it makes to the nucleus. When the user process dies, the envelope handles its legacy including closing any files that it had open and informing the Memory manager that the address space is defunct. If the user process's father is waiting for the death, the son's envelope wakes up the father's envelope. Once this clean up is finished, the envelope waits for a new user process to be born. It is the Family manager inside the User manager that handles relationships among processes as implied by the fork and wait system calls.

Most of the Tunis nucleus is machine independent. Its largest and most complex layer, the File manager, should not require any changes when porting the nucleus to a new machine.

THE ABSTRACTION OF ADDRESS SPACES

One of the key abstractions implemented by Tunis is that of an address space for each user process. This abstraction is implemented by the Memory manager using physical memory and disk space. Above the Memory manager the User and File managers assume that address spaces exist. They pass down each I/O request in terms of an *address descriptor*, consisting of a pair: (space number, virtual address). The User and File managers are not concerned with whether the users are swapped, paged or permanently memory resident. These concerns are isolated in the Memory manager. Below the Memory manager, the concepts of user address spaces and of user processes have disappeared. The individual device drivers deal only in physical memory addresses. As a result, a device driver is relatively easy to implement, because it does not depend on mechanisms such as swapping and paging. Nor does a device driver depend on details of the Unix file system. In brief, a device driver is neutral to its intended usage, and once written, could be used for other purposes besides supporting Tunis.

THE ASSASSIN PROCESS

When an interactive user of Unix types an "interrupt character" or a "quit character", the user processes associated with the terminal are notified, as if another process had executed a kill system call. The Teletype manager inside the Device manager layer detects these characters. It has no way to kill user processes because the concept of a user process does not exist at this level. The Teletype manager solves this problem by providing an entry called GetVictim. Whenever this is called, the Teletype manager returns the identity of a terminal that has received one of these characters.

This entry is called only from a process named the Assassin, which is inside the User manager. When the Assassin receives the terminal's identity, it sets a flag to notify each envelope whose user process is associated with the terminal. It is the envelope's responsibility to regularly check to see if his user process has been notified (assassinated). If so, the envelope takes the appropriate action, which often means destroying the user process.

Some system calls, such as a read from an inactive terminal, can block a user process for hours. When a process is killed during one of these long-lasting calls, it is not acceptable for the death to wait for the completion of the I/O. Tunis handles this difficulty by putting time limits on potentially long lasting I/O calls. When one of these calls does a time out, control is returned to the envelope in the User manager. This allows a test to see if the user process was assassinated. If not, the I/O request is reissued. If so, the envelope completes the assassination of the user process.

AN EXAMPLE MODULE

The implementation of Tunis will now be explored in more detail by considering one module, the Clock manager, in some depth. This module is chosen because it has an obvious function and a small implementation. A slightly simplified version of it has the following interface, as specified in CE. Its interface, specified by its CE stub, serves as a replacement for the actual Clock manager and allows separate compilation. It also contains comments that specify the actions of its exported procedures.

```
var Clock:
        external module
            exports(GetTime, SetTime, Delay)

            procedure GetTime(var t: LongInt) = external
                {Reads clock. Sets t to time in seconds}

            procedure SetTime(t: LongInt) = external
                {Changes clock. Set time to be t seconds}

            procedure Delay(t: LongInt) = external
                {Calling process is blocked for t seconds}

        end {Clock} module
```

The only externally visible parts of the Clock manager are its exported entry points GetTime, SetTime and Delay. An example of using the Clock manager is:

```
Clock.Delay(3)
```

This puts the caller to sleep for three seconds.

The actual implementation of the Clock manager has this form:

```
var Clock:
        module
            exports(GetTime, SetTime, Delay)

{1}         var ClockDoIO: {Machine dependent part}
                module
                    exports(WaitOneSecond)
                    procedure WaitOneSecond =
                        ...body of WaitOneSecond...
```

```
                    end module

{2}      var ClockMonitor:  {Encapsulates shared data}
              monitor
                    exports(GetTime, SetTime, Delay, Tick)

                    var now: LongInt := 0  {Value of clock}
                    ...other declarations...

                    procedure GetTime(var t: LongInt) =
                        imports(now)
                        begin
                            t := now
                        end SetTime

                    procedure SetTime(t: LongInt) =
                        ...body of SetTime...

                    procedure DeLay(t: LongInt) =
                        ...body of DeLay,
                            including wait statements...

                    procedure Tick =  {Called by Ticker process}
                        ...body for Tick, which updates "now", and
                            wakes up delayers...

              end monitor

{3}      {Entries to Clock module}
         procedure GetTime(var t: LongInt) =
              imports(var ClockMonitor)
              begin
                    ClockMonitor.GetTime(t)
              end GetTime

         ...analogous procedures SetTime and Delay...

{4}      {A process internal to the Clock manager}
         process Ticker
              imports(var ClockMonitor, var ClockDoIO)
              begin
                    loop
```

```
            ClockDoIO.WaitOneSecond
            ClockMonitor.Tick
        end loop
    end Ticker


    end {Clock} module
```

The Clock module contains four major parts, as numbered in the module's left margin. Many Tunis modules have this general form; the parts are:

1. A DoIO module. This contains machine dependencies and activates a particular device. In the case of the Clock manager, the device is the hardware clock. When Tunis is ported to a new machine the various DoIO modules in general must be re-written, because the devices of the new machine will require different logic to drive them. Each DoIO module should be textually isolated from its surroundings by placing its implementation in an include file; when porting Tunis, these include files are re-implemented.

2. A monitor. This contains data shared among processes. In the case of the Clock manager this data is the variable "now" which maintains the value of the clock. A monitor contains "conditions" to allow processes to wait for certain events, such as completion of an interval of sleep.

3. Module entries. These are the procedures and functions that are exported from the module and represent the interface to the Manager. In the case of the Clock manager, these entries simply route requests directly to the Clock monitor, which carries out the requests.

4. An internal process. There are typically one or more processes that call the monitor occasionally. These processes may be device drivers, as in the case of the Ticker process in the Clock Manager. In some cases, an internal process is an "ager"; for example the Cache manager has a process that repeatedly sleeps via "Clock.Delay" and then enters its monitor to consider re-allocation of the buffers in its pool. In other cases an internal process may be a courier or slave process; for example, the Memory manager may have a Swapper process that repeatedly enters its monitor to get a request for swapping and calls the Disk manager to perform the swap.

PROGRAMMING CONVENTIONS

The programming of the modules of Tunis is highly disciplined in that they are written in CE and they conform to the following conventions.

1. There are no interrupts. Interrupts are hidden in the kernel.

2. All I/O is synchronous with its invoking process. This means that a system process that activates a device waits until the device completes its requested action. Of course other processes are free to continue during the I/O.

3. System processes are never killed. All system processes in Tunis are created at system start-up and continue to exist until Tunis is shut down. The concept of dynamic creation and deletion of user processes is implemented by allocating and de-allocating envelope processes within the Family manager. Killing a user process causes its envelope to wait for a fork.

4. Unpredicted errors and exceptions in Tunis are fatal. However, predicted errors are handled by programming in CE. (Note that attempting to handle unpredicted error conditions in basic system software is at best extremely risky.)

These conventions make the programming of Tunis relatively straightforward. They disallow practices that make many systems difficult to understand and maintain.

ENTRY POINTS OF THE TUNIS KERNEL

In an ideal world the hardware processor directly supports the functions of Tunis's kernel. But in an actual implementation, the kernel is programmed mostly in assembler. The kernel provides the following functions:

1. Process synchronization. The CE compiler automatically generates calls to kernel entries for each entry to or exit from a monitor, for each signal and wait statement, and for the empty function that tests if a condition queue is empty. The compiler also generates calls to initialize descriptors for monitors and for condition queues.

2. I/O control. The kernel implements the BeginIO, WaitIO and EndIO routines, which support synchronous device operations. As explained in Chapter 4 in the section "Basic Device Management", a system process that wants to control a device first executes BeginIO to prevent premature interrupts from the device. Next it starts up the device and uses WaitIO to block itself until the device's interrupt occurs. During this waiting, other processes execute. When the interrupt occurs, the waiting process is dispatched with interrupts disabled. The process does any required post-

interrupt clean up and re-enables interrupts by executing EndIO.

3. User process control. The kernel supports an operation called RunUser that transfers control from an envelope process to its corresponding user process. This transfer implies a change in "protection domains", i.e., resetting the hardware protection registers. The user process continues to execute until it returns to the envelope via a trap. This return implies resetting the protection status to that of Tunis. There is a "time slice" parameter to RunUser that gives the maximum time the user process is to execute before returning to its envelope.

THE ENVELOPE AS GUARDIAN ANGEL

Each envelope process in the User manager successively ushers user processes through their lives from birth to death. When an envelope is in charge of a user process, it repeatedly executes the user process by calling the RunUser entry in the Memory manager.

The Memory manager sees that the user's address space is available to be executed; this may require swapping it into memory. Once this has been done, the Memory manager calls the RunUser entry in the kernel. This call changes the hardware memory mapping registers to locate the user's address space, loads the user's registers and status and executes the user process. When the user process traps, or when its time slice expires, the call to the kernel returns to the Memory manager. Eventually the Memory manager returns to the calling envelope. The envelope is given access to the trap number and parameters to the system call. In most systems, these parameters are passed as values in the user's general registers, which the envelope accesses.

Eventually, either the user process terminates itself by the exit system call or it is terminated by a "kill". When termination occurs, the envelope informs the Family manager via the procedure ReleaseUser. Here is a simplified version of the re-entrant procedure concurrently executed by all the envelope processes.

```
loop {For each user process to run, i.e., forever}
        Family.GetNewUser  {Birth of user process}

        ...

        loop {For each user process's request}

            Memory.RunUser  {User process runs}
            case trapNumber of
                getUId  => ...return user's Id.. end getUId
                kill  => ...set flag to kill process... end kill
```

```
hiccup => {End of time slice}
    if user was killed then
        exit {From request loop}
    end if
    end hiccup
processExit => {Suicide}
    exit {From request loop}
    end processExit
fork =>
    Family.Fork {Wake up another envelope}
    end fork
...other requests...
end case

end loop
...
Family.ReleaseUser {Death of user process}
end loop
```

From the envelope's viewpoint, the user process is a subroutine that is invoked in order to get a new request. From the user process's viewpoint, the envelope is a subroutine that carries out requests given in the form of system calls (traps). In reality, the two behave as a coroutine passing control back and forth as they switch between the system domain (Tunis) and the user domain (the user address space).

CHAPTER 9 SUMMARY

This chapter has overviewed the organization of Tunis. It is intended to be a high performance operating system nucleus that is compatible with Version 7 of Unix. It is intended to be portable and to have a clean internal structure.

Tunis has a layer structure, where the layers are as follows:

User manager. Contains the Envelope processes, which interpret user processes' system calls. The Envelopes call lower levels of the nucleus to carry out system calls. The User manager contains the Family Manager, which handles interactions among user processes.

File manager. Implements the Unix file system in a machine independent way. The File manager contains the Tree manager (supports the directory hierarchy), the Flat manager (supports flat files using i-nodes) and the Cache manager (contains buffers to minimize physical disk I/O). The Memory manager is called to access user processes address spaces.

Memory manager. Implements the concept of "address spaces" for user processes; locks down user memory as required for purposes such as DMA (direct memory access) from peripheral devices. The Memory manager does swapping for small systems such as the PDP-11 and paging for larger systems such as the VAX. The Memory manager translates virtual addresses within address spaces into physical addresses.

Device manager. Receives all requests for physical I/O and routes these requests to internal managers for disks, terminals, tapes, etc. These requests are in terms of physical addresses (not virtual addresses) because the concept of an address space does not exist at this level in Tunis.

Utility layers. Contains support software for higher layers. This software consists of the Clock manager, the Physical Memory manager and the Panic manager.

Kernel. Supports (1) process synchronization as implied by language features of CE, (2) I/O control via the Begin/Wait/EndIO operations, and (3) user process control via the RunUser operation.

The layers of Tunis are generally implemented as modules that contain monitors, sub-modules and processes. Machine dependencies are textually isolated; for example, a DoIO module contains the machine dependencies of each device manager. Processes (envelopes, agers, tickers, etc.) are used to manage concurrency.

CHAPTER 9 BIBLIOGRAPHY

Cardozo's thesis records work on the design and implementation of an earlier version of Tunis. Mendell's thesis records later work, with an emphasis on portability. Davis's thesis describes techniques for improving the reliability of Unix-like file systems.

Cardozo, P. Tunis-2: A Unix-like portable operating system, M.Sc. thesis, Computer Science Department, University of Toronto, 1980.

Davis, I.J. Towards reliable file systems. M.Sc. thesis, Department of Computer Science, University of Toronto, 1982.

Mendell, M.P. Toward portable operating systems. M.Sc. thesis, Department of Computer Science, University of Toronto, expected 1982.

CHAPTER 9 EXERCISES

1. In the Tunis nucleus, a system process is never killed. Describe the difficulties involved in killing system processes (e.g., when they are updating critical data).

2. Why does Tunis use an Assassin process (rather than having the terminal manager directly call the Family manager to kill user processes)?

3. Write a Clock manager in CE that meets the specification given in this chapter. Test it using a WaitOneSecond procedure whose body is the statement: Busy(1).

4. Write a simple Cache manager in CE. Its only entry point is:

    procedure Transfer(request: IORequest) = ...

IORequest is a record with fields specifying:

    Operation (read or write)
    Disk number
    Block number
    Substring (length and displacement from start of block)
    Address descriptor (space number and virtual address).

Each request to the Cache manager is for the transfer of a substring of a 256-byte block. The Cache manager contains an array of buffers holding recently accessed disk blocks. The Cache manager does disk I/O via the call Memory.Transfer. This passes an I/O request for a whole block down to the Memory manager. The Cache manager transfers a substring of a buffer to/from a user's address space via the call Memory.Move.

5. Some operating systems swap or page major parts of themselves. Discuss strategies for doing this to parts of the Tunis nucleus.

6. In Unix, tapes are (usually) considered to be "block devices" meaning that I/O is a block (256 bytes) at a time and that these blocks are buffered in a software cache. Discuss potential problems with this arrangement.

7. In Tunis, the details of process scheduling are hidden in the kernel and are not accessible to the memory manager. Discuss the advantages and disadvantages of this information hiding.

8. Consider the problem of printing out information about the internal state of the Unix or Tunis nucleus. There are two obvious ways to access the internal information. The first is to add new system calls that provide the information. The second is to read the nucleus's data using the sneak path provided by the special file that represents the nucleus's memory. (Unix uses this second strategy.) Gives the advantages of each of these two methods.

9. Some systems are structured by implementing each major layer as a process (rather than as a module as is done in Tunis). Contrast this structuring technique with the one used in Tunis.

10. List the abstractions or abstract machine operations supported by each of the major layers of Tunis.

Chapter 10

IMPLEMENTING A KERNEL

In previous chapters we have used processes and monitors without worrying about their implementation. We were content to treat them as well-defined abstractions, supported by the operating system, the language processor, or perhaps by some special hardware. This idea of an abstraction implemented by underlying software and hardware is one of the most powerful program structuring tools available, and is used constantly in software engineering. This concept has allowed us to concentrate on concurrency algorithms and operating system structures without the confusion of low-level, machine-dependent mechanisms such as interrupts.

But now we will confront the implementation problem, and take a close look at the "next lower level". Exactly how do we implement processes and concurrency constructs? The module (software or microprogram or hardware) that supports processes is called a *kernel*, and this chapter explains how to build a kernel. We give the design of a simple kernel for single CPU systems and show how this design can be implemented using production hardware (a PDP-11). We also give the design of a kernel for multiple CPU systems.

STRUCTURE OF A KERNEL

The main purpose of a kernel is to share CPU time among processes, so that each process has its own "virtual CPU". The kernel must also provide the process/process interface (to support interprocess communication) and the process/device interface (to support device management).

As the following diagram shows, the kernel receives traps from processes, dispatches processes (gives them the CPU), receives interrupts from devices, and issues Start I/O commands to devices. The kernel also receives interrupts from a clock.

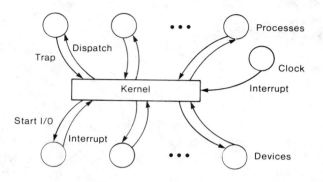

The kernel is the only module in the system that is aware of interrupts. All software activity outside the kernel is performed by processes, whose virtual CPUs are slowed down but otherwise unaffected by interrupts. In place of interrupts, the kernel provides special procedures that device manager processes use for controlling devices.

The kernel uses *process descriptors*, one per process, to keep track of the status of the processes and to allocate CPU time among them. When a process has not been allocated a CPU, its descriptor contains all the information needed to restart it. Typically, the descriptor will have fields for the process's instruction counter, general registers, floating point registers, addressing registers, and so on. The kernel dispatches a process (gives the process a CPU) by loading the CPU's registers from the process's descriptor. The loading of the CPU's instruction counter causes transfer of control to the process and the process resumes execution.

The kernel maintains a ready queue by linking together the descriptors of those processes that would be executing, except that a CPU is not available. The kernel also keeps track of blocked processes; to support monitors, it maintains a queue of process descriptors for each condition. A kernel might support any of the concurrency schemes described in Chapter 2, but in this chapter we will concentrate on the support of monitor features such as those in CE.

PROCESS/DEVICE COMMUNICATION

The process/device communication that must be supported by the kernel can be either synchronous or asynchronous. The synchronous method is conceptually simpler and is based on the DoIO operation. A device manager directs its device to perform a particular operation by executing

DoIO(deviceNo, command, status)

The status parameter is optional and can be used to indicate whether the operation was successful. The process executing DoIO is blocked until the requested I/O is complete.

DoIO can be implemented as an entry in the kernel; this was done in the Concurrent Pascal system. After the kernel starts up the device, using the given command, the kernel can allocate the CPU to another process. When the device completes the operation, it notifies the kernel (by an interrupt) and the kernel can again dispatch the device manager.

Chapter 4 showed how DoIO for each device can be implemented using three kernel operations: BeginIO, WaitIO and EndIO. The advantage of providing these three operations rather than supplying DoIO as a kernel entry is that new devices can be handled without modifying the kernel.

The DoIO operation is conceptually simple because it makes the device manager process synchronous with its device. The device is active only when the manager is blocked waiting for the DoIO operation. Conceptually, the manager and the device are one process, executing part of the time as a software process (the manager) and part of the time as a hardware process (the device).

The problem with DoIO is that the manager may need to continue execution while the I/O is in progress. For example, the manager may need to accept further user I/O requests and sort these requests by priority before the device completes. This problem can always be solved by splitting the manager into two processes. The first uses DoIO to control the device and the second receives users' requests and transfers them to the first. This is a good solution when using CE, because its processes are quite efficient. But if processes and interprocess communication are costly, an alternative to DoIO may be required.

The alternative method allows the device and its manager process to be asynchronous. It uses the following commands.

SignalDevice(deviceNo, command)
WaitDevice(deviceNo, status)

Executed one immediately after the other, these two commands are equivalent to DoIO. When executed separately, they allow the manager process to continue execution during the I/O. A variation of WaitDevice could allow the manager to test the device to see if the operation is completed, without being forced to wait for the completion. Signal/WaitDevice is somewhat more costly than DoIO to implement, because the kernel is forced to maintain a descriptor to record the status of each device. With DoIO, these descriptors can be avoided because the status of a completing device can be immediately transferred to the device's waiting manager. In this chapter we will ignore Signal/WaitDevice in favor of the simpler DoIO.

QUEUE MANAGEMENT

The kernel manages several queues of process descriptors, including the ready queue. To support monitors, there must also be a queue for each of the conditions. We will use the following notation for queue removal and insertion operations:

> Remove item from queue
> Insert item into queue

For process queues, each item can be located by a pointer to its process descriptor. The queues in the kernel are typically priority ordered or FIFO (first-in-first-out). Either method of ordering can be accomplished by very simple programs, which we will now develop. Persons well acquainted with queuing algorithms may choose to skip the rest of this section.

FIFO queues can be represented by pointers called first and last and a field named next in each item, as shown here.

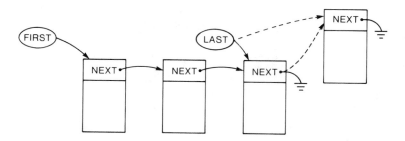

The dotted arrows show how a new item is inserted at the end of the queue. When the FIFO queue is empty, it has this form:

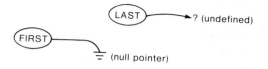

The test to see if the queue is empty is accomplished by comparing the value of first to the null pointer. Here are the programs for removing and inserting nodes in FIFO queues:

Remove item from queue:
    item := first
    first := node(first).next

Insert item into queue:
    if first = null then {Is queue empty?}
        first := item
    else
        node(last).next := item
    end if
    node(item).next := null
    last := item

The operations for priority queues are only slightly more complicated, and we will now develop them.

Priority queues can be represented by a pointer called first together with fields called next and prty in each item, as shown here:

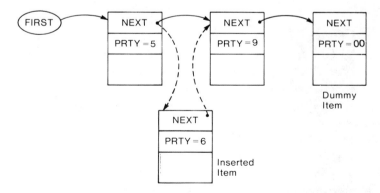

We are assuming that low-numbered priorities come first in the queue. The dotted arrows show how a new item with priority 6 would be inserted between items with priorities 5 and 9. The dummy item has a priority that is larger than any real item; its purpose is to simplify insertion, so there is no special case for inserting at the tail (just before the dummy). The test to see if the queue is empty is accomplished by comparing the value of first to the pointer to the dummy item. Here are the programs for removing and inserting items.

Remove item from queue:
     item := first
     first := node(first).next

Insert item into queue:
     if node(item).prty < node(first).prty then
          {Insert item first in queue}
          node(item).next := first
          first := item
     else   {Find place to insert item}
          previous := first
          loop
               link := node(previous).next
               exit when node(item).prty < node(link).prty
               previous := link
          end loop
          node(previous).next := item
          node(item).next := link
     end if

In a following section we describe how a kernel can use queues to support monitors.

ENTRIES INTO THE KERNEL

The kernel can be entered due to action on the part of a process, a device or a clock, as was shown in the diagram of the kernel's structure. These are the types of entries:

Traps - calls from processes, requesting operations such as the wait statement.

Device interrupts - signaling that requested operations are complete.

Clock interrupts - signaling the end of a process's time slice.

To implement a kernel, we will need a program for each of these entries. Given that the kernel supports monitors, we also need entries for:

EnterMonitor(m): The process is entering monitor m and exclusive access to m must be guaranteed.

ExitMonitor(m): The process is leaving monitor m, and another process may be allowed to enter m.

Signal(c): The process is inside a monitor and signals condition c; if a process is waiting for c, it must be awakened.

Wait(c): The process is inside a monitor and waits for condition c; this releases the monitor. The CPU can be allocated to another process.

DoIO(deviceNo, command, status): The command is to be passed to the specified device, and the executing process is to be blocked until the device completes its operation; this releases the CPU so it can be re-allocated.

Rather than directly supporting DoIO, the kernel may have entries for Begin/Wait/EndIO. If monitors are embedded in a high-level language, it is the responsibility of the compiler to generate code to invoke the entries. If the programmer is using a language that does not support monitors, for example, assembly language, he can still use monitors by using macros or procedures that invoke the kernel.

Before giving programs to implement these entries, we will discuss some assumptions that can make implementation of monitors easier.

SIMPLIFYING ASSUMPTIONS

A kernel can be small and fast, or large and inefficient, depending on the hardware's interrupt structure and addressing mechanisms. The complexity of the kernel also depends on how much support it must provide for the operating system's memory allocation, accounting and protection policies. It also depends on details of process/process and process/device communication; for example, message passing is inherently more complex than semaphores.

In this chapter we will largely ignore the problems of memory allocation, accounting and protection. Since those problems are handled simply in many small operating systems, the kernel designs we will give are directly applicable to such systems. Simple solutions to these problems are also typically used in the inner layers of large operating systems; in these our designs are of immediate use to support activities such as device management.

Previous chapters have explained how more elaborate facilities, such as swapping of user processes, can be implemented using the CE language's concurrency features.

The architecture that the hardware provides for controlling devices can make or break an operating system. If these mechanisms, primarily the Start I/O instruction and device interrupts, are unstructured and imply extensive interactions among channels and devices, no reasonable structure may be possible. In such a case it will be hopeless to try to impose the elegance of a mechanism such as DoIO. We will assume for our kernel implementation that the architecture is relatively clean.

A KERNEL FOR SINGLE CPU SYSTEMS

A kernel that supports multiple processes, monitor entry and exit, and signal/wait can be quite small. For example, the kernel developed in a following section for a PDP-11 consists of less than 50 machine instructions. In the present section we will give a detailed design for a simple kernel that could be implemented on typical production computer systems. We will assume that there is only one CPU because this makes the design simpler, but in a later section we will show how to handle multiple CPUs.

The reason a single CPU system is particularly easy to handle is that mutual exclusion can be guaranteed simply by disabling interrupts. Taking this idea to extremes leads to the monolithic monitor, as discussed in Chapter 1, in which all operating system activities are performed with interrupts disabled. By contrast, our first design will disable interrupts only while the kernel or a monitor is active. This design will impose the restriction that at most one monitor can be active at any given time.

The following diagram illustrates the queues of process descriptors that the kernel manages. The diagram shows one running process, three ready processes, two processes waiting for condition cond1, none for cond2, one for cond3, none for device1, and one for device2.

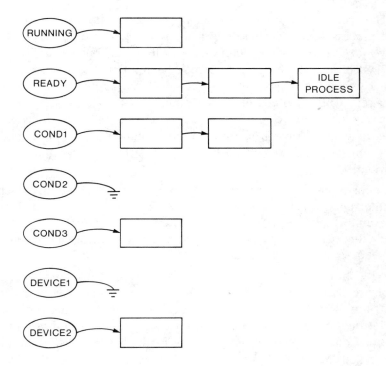

Each rectangle in this diagram represents a process descriptor. The links to descriptors can be represented as machine addresses. If priority-ordered queues are used, the null pointer can correspond to the address of the dummy descriptor.

When all processes are blocked, the CPU becomes idle. This special situation can be handled in an elegant way by using an "idle process", which has a lower priority (larger number) than any other process. In the diagram the idle process is shown last on the ready queue. When all "real" processes become blocked, the idle process is dispatched (becomes the running process). Any "real" process that becomes ready will immediately be given the CPU because its priority is greater than the idle process. With this arrangement, the kernel needs no special logic to handle the case of the idle CPU. If desired, the idle process can absorb CPU time by executing a non-urgent program, such as computing the digits of the irrational number pi. Note: the priority of the idle process must be less urgent than (a smaller number than) the priority in the dummy descriptor, which always marks the end of the ready queue.

The speed of the kernel depends largely on the time taken to save and restore process status. This saving and restoring is called *task switching* or *process switching.* Some computers have facilities for process switching in a single instruction; in others, there may be multiple register sets and the saving or restoring is accomplished simply by changing the pointer to the register set. Unfortunately, on some computers, status saving and restoring is relatively slow. For example, on the System 360/370, the process status consists of over 100 bytes of information (program status word, float and general registers), and several instructions must be executed to do the process switch.

In many computer systems, the EnterMonitor operation, which disables interrupts, can be implemented by a single machine instruction. We will assume that a special register determines whether interrupts are enabled; this register is part of a process's status and must be saved in the process's descriptor.

With these preliminaries out of the way we now give the program for each of the entries into the kernel. In these programs, we use a pointer called "running" to locate the process descriptor of the running process. For example, the line "Remove running from ready" means to remove a process descriptor from the ready queue and to place the descriptor's location in the pointer called "running".

EnterMonitor: Disable interrupts

ExitMonitor: if first ready priority < running priority then
                    Save status of running
                    Insert running into ready
                    Remove running from ready
                    Restore status of running
                end if
                Enable interrupts

Wait(c):   Save status of running
                Insert running into c
                Remove running from ready
                Restore status of running

Signal(c):   if not empty(c) then
                    {Give CPU to waiting process}
                    Save status of running
                    Insert running into ready
                    Remove running from c

> Restore status of running {Interrupts stay disabled}
> end if

Slice: {Clock interrupt}
              Save status of running then disable interrupts
              Insert running into ready
              Remove running from ready
              Restore status of running {Enables interrupts}

The program for ExitMonitor checks to see if the CPU should be given to another process (this would be a process that did a signal). If so, the program does a process switch, transferring the CPU to the other process.

The program for Wait saves the running process's status, puts its descriptor on the queue for condition c, takes another process's descriptor from the ready queue, and dispatches that process. The line "Save status of running" uses the pointer called running to locate a process descriptor, and saves the current status of the CPU in that descriptor. The line "Insert running into c" places this pointer in the queue for condition c. The line "Remove running from ready" changes the pointer called running so it locates another process descriptor. The line "Restore status of running" includes resetting the instruction pointer and thus transferring control to the new process. In our simple monitor, Wait and Signal can only be executed inside a monitor; we know interrupts have been disabled by Enter-Monitor. This disabling guarantees mutually exclusive access to the queues of processes.

The program for Signal does nothing unless it finds a process waiting for the condition. If one is found, the running process has its status saved and its descriptor placed on the ready queue. Then a waiting process's descriptor is removed from the condition queue and that process is dispatched.

A time slice allows another process to run. We are assuming that the ready queue is priority-ordered, and that within a given priority, it is FIFO; this corresponds to the implementation of priority queues we have given earlier in this chapter. The program for Slice saves the status of the running process and puts its descriptor on the ready queue; then it takes a descriptor from the ready queue and dispatches that process. If all processes have the same priority, this results in round-robin scheduling. If the interrupted process has a higher priority than any ready process, the Slice program causes the interrupted process to be re-dispatched. In this last case, status saving and restoring could be avoided by first comparing the running priority to the first ready priority, and immediately redispatching if appropriate. An expanded version of the Slice program with this extra test would be more efficient in computers that have slow process switching.

Our kernel allows a low priority process in the ready queue to be indefinitely overtaken by higher priority processes. If this situation can arise (for other than the idle process) in a particular system, it can be avoided by having the kernel dynamically decrease the priority of processes that use a lot of CPU time.

We have now given all of the kernel except the part that allows processes to control devices.

HANDLING INPUT AND OUTPUT

We can augment our kernel to handle input and output by giving it an entry point called DoIO and a handler for input/output interrupts. We will assume that processes can call DoIO only from outside monitors. The implementations are as follows.

DoIO(deviceNo, command):
    Save status of running then disable interrupts
    Insert running into queue(deviceNo)
    StartIO(deviceNo, command) {Hardware instruction}
    Remove running from ready
    Restore status of running

IODone(deviceNo): {Handle I/O interrupt}
    Save status of running then disable interrupts
    Insert running into ready
    Remove running from queue(deviceNo)
    Restore status of running {Enables interrupts}

The DoIO operation causes the kernel to place the executing process on a queue waiting for the device to finish, start up the device, and then find a new process to dispatch. We will assume that only one process will request I/O for a particular device. That process will be the device's manager and may receive I/O requests from other processes. There are a number of advantages of using a separate manager process, such as providing I/O buffering and protection.

The IODone routine handles I/O interrupts. First it saves the status of the running process and puts its descriptor on the ready queue. Then it removes the waiting process from the device's queue and dispatches that process. The IODone routine could be made a bit more sophisticated by checking to see if the waiting process has a higher priority than the interrupted process. On the other hand, if all device manager processes are relatively simple and fast, IODone as given is probably best because it allows the manager to re-start the device immediately. If there is device status associated with the I/O interrupt, it would be passed by the kernel to the managing process in an extra parameter of DoIO.

The Begin/Wait/EndIO operations can be implemented in an analogous fashion. Assuming these are used only outside of monitors, BeginIO and EndIO can be implemented the same way as EnterMonitor and ExitMonitor, respectively. WaitIO can be implemented the same as a Wait on a condition queue for the device. Interrupts are handled exactly as in IODone above.

The kernel design we have presented implies the restriction that monitors cannot call each other. If this kernel were used to support CE, we would have to impose the same restriction on CE. Alternatively, we could extend our kernel design to handle the more general case of monitors calling monitors, which CE allows. We will not go into this extension, but will point out that the multi-CPU kernel given later in this chapter can handle a single CPU in which monitors call each other.

Unfortunately, some current computer architectures have a relatively complex I/O structure, with external channels, controllers and devices that must be managed from the CPU. The elegant handling of I/O that we have just developed may not be directly adaptable to those systems, especially if an error on a device can be reported by an interrupt separate from the one for device completion. Since the PDP-11 architecture is clean and widely used, we will show how our kernel design can be translated into its machine language. Persons who are not familiar with machine language may choose to skip the following section.

A KERNEL FOR THE PDP-11

The two preceding sections have given the design of a kernel that supports monitors, time-slicing and process/device communication for a single CPU system. We can implement a kernel for a particular computer by translating this design into machine instructions. For the PDP-11 this translation is particularly simple, and we will now give it.

On the PDP-11, process status consists of a processor status word (PS), a program counter (PC), a stack pointer (SP), and six general purpose registers called R0, R1, up to R5. All these must be saved when the process is interrupted. Each process has a stack (a contiguous allocation of memory) whose currently last active word is pointed to by its SP register.

The PS contains a hardware priority with values from 0 to 7. This priority is not the priority used in ordering the ready queue. Rather, the hardware priority determines the disabling of device interrupts. When the PS priority is 7, no device can interrupt. When it is n, only devices with a hardware priority greater than n can interrupt. For our kernel we will use this disabling mechanism in the simplest possible way: we will use only priority 7 (to disable all interrupts) and priority 0 (to enable all interrupts). Whenever an interrupt or trap occurs, the PDP-11 hardware stores the current PS and PC on the current stack and loads a new PS and PC corresponding to the interrupt or trap. The remainder of a process's status (SP and six registers) can then be saved by the kernel.

We will assume that the kernel maintains a descriptor for each process that has the following fields:

NEXT	- pointer to next process descriptor
PRTY	- priority for ready queue
SAVER1	- saves status of R1
...	
SAVER5	- saves status of R5
SAVESP	- saves status of SP

Since the PS and PC are automatically saved on the process's stack, they do not require fields in the process descriptor.

We will assume that whenever a process is running, register R0 is pointing to the process's descriptor. We will use the mnemonic RRUN as a synonymn for R0. Following an interrupt, these following instructions complete the saving of the interrupted process's status:

```
MOV     R1,SAVER1(RRUN);
MOV     R2,SAVER2(RRUN);
...
MOV     R5,SAVER5(RRUN);
MOV     SP,SAVESP(RRUN);
```

When a process is to be dispatched, these registers can be restored by loading R0, i.e., RRUN, with the address of the appropriate descriptor and executing:

```
MOV     SAVER1(RRUN),R1;
MOV     SAVER2(RRUN),R2;
...
MOV     SAVER5(RRUN),R5;
MOV     SAVESP(RRUN),SP;
RTI;
```

The RTI (return from interrupt) instruction restores the PC and PS from the top of the process's stack, and the process resumes execution. Note that the descriptor does not require a field for R0 (RRUN) because the link field pointing to the descriptor will hold R0's value.

The following diagram gives the descriptors of a system having five processes and a single condition queue.

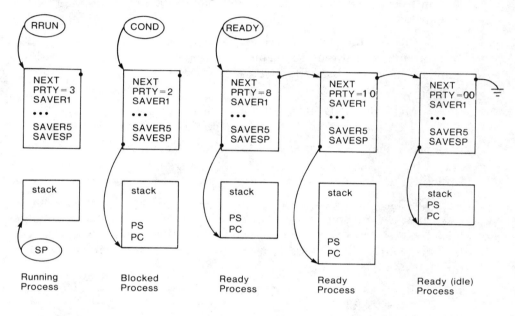

| Running
Process | Blocked
Process | Ready
Process | Ready
Process | Ready (idle)
Process |

As shown, one process is running and another is waiting for a condition. Three processes, including the idle process, are in the ready queue.

There are two assumptions we will make to simplify our PDP-11 kernel. First, we will assume that a process calls the kernel by executing a trap instruction, and when it does so, it is not currently using registers R1 through R5. This assumption eliminates the need to save and restore these registers before and after traps. Second, we will assume that each condition will have at most one process waiting for it. This second assumption is too severe for many systems including CE, but it simplifies our kernel. It is always possible to program using only single process conditions by using arrays of conditions; this must be done in the language Concurrent Pascal which supports only single process condition queues. If required, our implementation can be expanded to handle the more general case.

With these preliminaries out of the way, we will give the actual code for the PDP-11 kernel. Interrupts can be disabled by a single instruction that sets the PS priority to 7, as follows.

```
EnterMonitor:        BIS #PBITS,PS; Set priority bits to ones
```

The BIS (bit set) instruction sets the three-bit hardware priority field to ones (octal 7). This instruction can be emitted in-line whenever a monitor is to be entered. Analogously, the following instruction enables interrupts:

```
BIC #PBITS,PS; Clear priority bits to zeroes
```

This instruction is used in the implementation of ExitMonitor, which is given below.

The code to implement Wait and Signal is a bit longer, so we will factor it out into kernel routines that are called by traps. Each call to the Wait routine is done by this in-line code:

```
Call Wait:
    MOV  #COND,RCOND;  Put number of condition in a register
    TRAP #WAITNO;      Trap, passing number of wait routine
```

A similar calling sequence is used for Signal. Recall that we are assuming that at the call, registers R1 to R5 (except RCOND) are not in use, so the kernel will not need to save them.

The trap causes the hardware to save the process's PC and PS and to pick up a new PC and PS from a fixed location in low memory. The new PS will be set up to have priority 7, so interrupts will remain disabled. The new PC points to the following segment of code, called ROUTER, that in turn transfers control to the appropriate trap handler.

```
ROUTER:
    MOV  (SP),RTEMP;            Locate trap instruction
    MOVB -1(RTEMP),RTEMP;       Load trap no. from instruction
    JMP  @VECTOR(RTEMP);        Jump to correct handler
```

The jump to the correct trap handler is done indirectly using a "transfer vector", which is a sequence of words in memory that holds the addresses of the handlers.

The Wait and Signal routines, which gain control from ROUTER, are pleasingly small: only 11 instructions for Wait and 9 instructions for Signal. The Signal routine includes a call to an INSERT routine (16 instructions) that inserts a process descriptor into the priority-ordered ready queue. The INSERT routine is also called by the SLICE routine. Here is the code for these routines.

```
WAIT:
    MOV  SP,SAVESP(RRUN);       Saves status of running
    MOV  RRUN,(RCOND);          Insert running into C
DISPCH:
    MOV  READY,RRUN;            Remove running
```

```
        MOV    NEXT(RRUN),READY;           from ready
        MOV    SAVER1(RRUN),R1;        Restore registers of running
        MOV    SAVER2(RRUN),R2;        Restore registers of running
        MOV    SAVER3(RRUN),R3;        Restore registers of running
        MOV    SAVER4(RRUN),R4;        Restore registers of running
        MOV    SAVER5(RRUN),R5;        Restore registers of running
        MOV    SAVESP(RRUN),SP;        Restore registers of running
        RTI;                           Restore PC and PS
SIGNAL:
        TST    (RCOND);                if empty(C) then
        BNE    SWAP;                       Cause return to signaler
        RTI;                           else
SWAP:
        MOV    SP,SAVESP(RRUN);            Save status of running
        JSR    PC,INSERT;                 Insert running into ready
        MOV    (RCOND),RRUN;              Remove running
        CLR    (RCOND);                      from C
        MOV    SAVESP(RRUN),SP;           Restore status of running
        RTI;                           end if {Restore PC, PS}
SLICE:
        MOV    R1,SAVER1(RRUN);        Save registers of running
        MOV    R2,SAVER2(RRUN);        Save registers of running
        MOV    R3,SAVER3(RRUN);        Save registers of running
        MOV    R4,SAVER4(RRUN);        Save registers of running
        MOV    R5,SAVER5(RRUN);        Save registers of running
        MOV    SP,SAVESP(RRUN);        Save registers of running
        JSR    PC,INSERT;              Insert running into ready
        BR     DISPCH;                 Remove running from ready and
        ;                                  dispatch running
INSERT:
        MOV    READY,RLINK;            link:=ready
        CMP    PRTY(RRUN),PRTY(RLINK); if prty(running) > prty(link)
        BLE    MERGE;                  then
        MOV    RLINK,NEXT(RRUN);       next(running):=link
        MOV    RRUN,READY;             ready:=running
        RTS    PC                      else
MERGE:
        MOV    RLINK,RPREV;            prev:=link
        MOV    NEXT(RLINK),RLINK;      link:=next(link)
LOOP:                                  loop
        CMP    PRTY(RRUN),PRTY(RLINK); exit when prty(running)
        BGT    HOOKUP;                        > prty(link)
```

```
    MOV   RLINK,RPREV;              prev:=link
    MOV   NEXT(RLINK),RLINK;        link:=next(link)
    BR    LOOP;                     end loop
HOOKUP:
    MOV   RRUN,NEXT(RPREV);         next(prev):=running
    MOV   RLINK,NEXT(RRUN);         next(running):=link
    RTS   PC;                       end if
```

We can implement ExitMonitor as a conditional trap to SLICE followed by enabling interrupts:

```
ExitMonitor:
    MOV   READY,RREADY;             if prty(ready)<prty(running)
    CMP   PRTY(RREADY),PRTY(RRUN);
    BGE   CONTINUE;                 then

    TRAP  #SLICENO;                 Invoke SLICE in kernel
CONTINUE:                           end if
    BIC   #PBITS,PS;                Enable interrupts
```

This code can be emitted in-line whenever a process returns from a monitor. We will not give the routines for handling I/O. The DoIO routine can be modeled after Wait and the I/O interrupt handler can be modeled after SLICE. The PDP-11 kernel we have given can be expanded in many ways; in the exercises we suggest a number of improvements and expansions.

A KERNEL FOR MULTIPLE CPU SYSTEMS

We have given a simple, efficient kernel that is based on traditional, interrupt-oriented, single CPU architecture. In this section we will give the design of a kernel that implements processes and supports monitors when there are several CPUs. The implementation will be quite small, but requires some study to be well understood.

We will present the design in terms of an abstraction (a virtual machine) which is not directly supported by currently available hardware. Then we will show how this abstraction can be implemented by a few machine instructions.

Our abstraction allows the logic of the kernel to be quite elegant, because it lets us assume that the kernel runs on its own machine (a virtual processor). Since this processor executes only the kernel and it does not accept interrupts, mutually exclusive access to queues of descriptors is guaranteed. Whenever a process wishes to execute a primitive, such as Wait, the process's request is delayed until the kernel is idle; then the kernel is activated by invoking its appropriate routine, and is handed a pointer to the calling process's descriptor. It is assumed that the calling process has stopped running and its status has been saved. Once activated, the kernel is free to manipulate whatever queues it maintains.

The kernel's ready queue gains a very special significance. Whenever the kernel inserts a process descriptor into the ready queue, the kernel's virtual processor allocates a CPU to the process. This convenient assumption means the kernel does not need to worry about the mechanics of dispatching processes or sharing CPUs. When a running process calls the kernel or is interrupted, the process is no longer on the ready queue and the virtual processor has saved the process's status. The kernel is concerned only with making processes ready (by putting them on the ready queue). Later we will show how these convenient assumptions are supported.

To implement monitors, the kernel has a queue for each condition. It also has an entry queue for each monitor, and allows more than one monitor to be active at once. Of course, at any given time it does not allow more than one process to be active in a particular monitor. Associated with each monitor is a flag called occupied which records whether the monitor is currently busy.

Given these data structures and our delightfully powerful virtual processor, the kernel can be implemented in the following few lines.

```
EnterMonitor(m):
    if occupied(m) then
        Insert caller into m
    else
        occupied(m) := true
        Insert caller into ready
    end if


ExitMonitor(m):
    if empty(m) then
        occupied(m) := false
    else {Transfer process from m's queue to ready queue}
        Remove p from m
        Insert p into ready
    end if
    Insert caller into ready


Wait(c):
    if empty(m) then
        occupied(m) := false
```

```
        else {Transfer process from m's queue to ready queue}
            Remove p from m
            Insert p into ready
        end if
        Insert caller into c

    Signal (c):
        if empty (c) then
            Insert caller into ready
        else {Transfer process from c's queue to ready queue}
            Remove p from c
            Insert p into ready
            Insert caller into m
        end if
```

The EnterMonitor routine tests to see if the monitor is busy. If so the calling process, whose descriptor is pointed to by "caller", is placed on a queue waiting for the monitor to be free. If not, the busy flag is set to true, and the process is re-activated by placing its descriptor on the ready queue. The other routines are straightforward and we will not discuss them.

The handling of I/O is also straightforward, and can be implemented as follows:

```
    DoIO (deviceNo, command):
        StartIO (deviceNo, command) {Actually start the device}
        Insert caller into queue (deviceNo)

    IODone (deviceNo): {I/O interrupt handler}
        Remove p from queue (deviceNo)
        Insert p into ready
```

The Start I/O instruction activates the device, and the interrupt signals that the device is again idle. One of the beauties of this arrangement is that processes and devices are very similar from the point of view of the kernel. The analog of the Start I/O for a device is the insertion of a process descriptor into the ready queue. The analog of a device interrupt is a process trap. The interrupt (trap) tells the kernel that the device (process) requires attention. The device (process) remains idle until re-started by the kernel. The device (process) is restarted by a Start I/O instruction (by insertion into the ready queue).

SUPPORTING THE KERNEL'S VIRTUAL PROCESSOR

We have now designed the multiple CPU kernel, but we have not yet shown how to support its virtual processor. We might ask a hardware designer or a microprogrammer to implement the special processor. There would be a bonus to such an implementation: it could as well be used to implement other synchronization schemes such as semaphores or message passing. So, the virtual processor would be useful on its own right, independent of monitors. If special hardware or microprogramming are not available, we can still easily and efficiently implement the kernel's processor, as we will now show.

We will make the following assumptions about the system's architecture. There are n identical CPUs, and each CPU has its own interrupting clock. We will use n idle processes to sop up idle CPU time. All the CPUs can address the same memory, so each CPU can run any process and each CPU can access the kernel's data structures. Each device can interrupt any CPU. The hardware is constructed such that the interrupted CPU is the one currently running with the lowest (highest numbered) priority. There is a special instruction called Preempt CPU, which any CPU can execute; it interrupts another CPU (the one running the lowest priority process).

We will not dedicate a CPU to the kernel (that would be an awful waste). Instead, we give the kernel the appearance of having a dedicated CPU by temporarily borrowing whatever CPU received the interrupt (or trap).

The following three code segments implement the special virtual processor. They share the n CPUs among the processes that have been put on the ready queue, giving CPUs to the n highest priority processes. These segments make sure that only one CPU at a time enters the kernel. They use two routines called Enter Kernel and Exit Kernel to save process status, gain mutual exclusion and dispatch processes. These two routines will be shown later.

```
Trap (trapNo): {Process invokes the kernel}
        EnterKernel {Save status and gain mutual exclusion}
        TrapHandler (trapNo,running) {Handle trap, such as
            EnterMonitor, and return here. The handler may
            insert the process into ready}
        ExitKernel {Release mutual exclusion and dispatch}


Slice: {CPU interrupted by clock or by PreemptCPU}
        EnterKernel {Save status and gain mutual exclusion}
        Insert running into ready
        ExitKernel {Release mutual exclusion and dispatch}
```

IOInterrupt(deviceNo): {CPU interrupted by device}
    EnterKernel {Save status and gain mutual exclusion}
    Insert running into ready
    Call IODone(deviceNo) {Wakes up device manager process}
    ExitKernel {Release mutual exclusion and dispatch}

For these routines, each CPU must have private variables (these could be registers) called trapNo, running, and deviceNo.

A process invokes the kernel (for example, to enter a monitor) by executing Trap. This saves the process's status and passes a pointer to the process's descriptor to the appropriate trap handler. The process's descriptor is no longer on the ready queue, so the process will remain blocked until the kernel inserts the descriptor into the ready queue. For example, ExitMonitor, which is a trap handler, always inserts the calling process's descriptor into the ready queue.

Slice is invoked when a CPU's clock causes an interrupt. The interrupted process's status is saved and its descriptor re-inserted into the ready queue. As was the case for the single CPU kernel, this results in round-robin scheduling within priorities.

IOInterrupt is like Slice but additionally removes the device manager's descriptor from the device queue and inserts it into the ready queue. As a side effect, this reschedules the executing process behind other processes of the same priority. If this rescheduling is considered undesirable, then the interrupted process can be inserted first in the ready queue.

IMPLEMENTING KERNEL ENTER/EXIT

On a single CPU system, we can implement EnterKernel by simply saving the process's status and disabling interrupts. ExitKernel removes the highest priority process from the ready queue, restores its status and enables interrupts. Here are these two operations:

    EnterKernel: {For single CPU}
    Save status of running and disable interrupts

    ExitKernel: {For single CPU}
    Remove running from ready
    Restore status of running {This enables interrupts}

These operations can be made faster for particular computer architectures by such tricks as avoiding saving a process's status when it is executing a non-blocking trap. With these implementations of Enter/ExitKernel, the kernel we have just given can be used on a single CPU system. This single

CPU kernel has the advantage that it allows more than one monitor to be active at once and allows a monitor to call another monitor (this is not possible with the previous single CPU kernel). The disadvantage of this new single CPU kernel is that it is somewhat more complex and slower than our previous design.

In a multiple CPU system we must guarantee that only one of the several CPUs is allowed into the kernel at a given time. This can be done by a test-and-set loop, which was described in Chapter 2. For the multiple CPU system, the following instruction sequences implement Enter/ExitKernel.

```
EnterKernel:
    Save status of running and disable interrupts
    loop
        TestAndSet(kernelOccupied, wasOccupied)
        exit when not wasOccupied  {Test old flag value}
    end loop
```

```
ExitKernel:
    Remove running from ready
    needCPU :=
        (first ready priority < lowest running priority)
    kernelOccupied := false  {Release control of kernel}
    if needCPU then  {Important process was readied}
        PreemptCPU
    end if
    Restore status of running  {This enables interrupts}
```

EnterKernel begins by saving the process's status. Then it disables interrupts, which effectively captures the process's CPU, so it cannot be pre-empted by an interrupt. Finally, this CPU is used to execute a test-and-set loop, which eventually gains control of the kernel. ExitKernel is essentially the inverse of this: the kernel is released by resetting kernelOccupied to false, the status of the new running process is restored and interrupts are enabled thereby allowing pre-emption of the CPU. A slight complication arises from the fact that the processes that are waked up may have higher priorities than processes already running on CPUs. When this occurs, the PreemptCPU instruction is used to force another CPU to execute Slice.

We have now given code segments that implement the virtual processor required by the kernel, and we have given the code segments that comprise the kernel. Together these support monitors for a multiple CPU computer system. Such a system has obvious advantages in terms of reliability and expandability. Reliability is improved because any failing CPU can simply be retired with its current process transferred to the ready queue. (Hopefully, the failing CPU does not destroy any critical data such as the kernel's queues.) Since all CPUs are the same, loss of any particular CPU only slows down the system without causing a disaster. Expandability is improved because when the system needs more computing power, another CPU can be added without affecting the software. Adding or removing a CPU changes system throughput, but does not affect the system's correctness.

KERNELS FOR CE AND TUNIS

The first kernel for CE was constructed in assembly language for the MC68000. This kernel is similar to the multiple CPU kernel described in this chapter. The MC68000 kernel was later re-written to support other computers, notably the PDP-11 and the MC6809 microprocessor.

Writing one of these kernels is straightforward, but it requires considerable care to avoid bugs. The problem is that the handling of interrupts and process switching is inherently complex, and assembly language is not an easy language to deal with.

CE programs can be run under Unix or other operating systems. This is done by sharing the time of a Unix user process among the CE processes. This sharing is accomplished by a simulation kernel, which does the CE process switching. Unix remains unaware of the fact that a single user process is made to behave like multiple CE processes. CE's simulation kernel is very similar to its corresponding bare machine kernel. One of the differences is that the simulation kernel does not use a clock device to do time slicing. Instead, the CE program includes calls to the kernel to tell the kernel when to do time slicing. These calls are implicitly emitted by the compiler for CE programs to be run in simulation mode. If these calls are not emitted, the simulation kernel still works. However, without them, it cannot regularly gain control from a long running process, and so it cannot do fair scheduling of the available CPU time.

The kernel for Tunis is a CE kernel that supports Begin/Wait/EndIO instead of DoIO. The Tunis kernel has entries that increase and decrease a process's software priority. These priorities are used to make some CE processes in the Tunis nucleus run faster to improve Tunis's performance; however, they do not affect correctness. In particular, they are *not* used to gain mutual exclusion; monitors are used for mutual exclusion. The Tunis kernel also has entries to associate an envelope process (a system process) with its corresponding user process. The kernel is responsible for changing the memory protection registers when an envelope becomes a user process and vice versa.

The first chapter of this book described the techniques of basing an operating system on a kernel. The main responsibilities of the kernel were defined as the support of processes along with process/process and process/device communication. From Chapter 1 until the present chapter, we used processes without being concerned about how they were supported. In this chapter we have returned to the subject of kernels, and have shown how the concurrent algorithms presented throughout this book can be supported by a kernel using traditional computer hardware.

CHAPTER 10 SUMMARY

In this chapter we have seen that a kernel is entered when the following occur:

Trap - a process requests a service. When the service has been provided, the process can be dispatched (given a CPU).

I/O interrupt - a device completes an operation. When the device is to be restarted, a Start I/O instruction is executed.

Clock interrupt - may signal the end of a process's time slice.

The traps are used to invoke primitive operations; to support monitors, the kernel may have entries for traps corresponding to:

Entering a monitor.
Exiting a monitor.
Signaling a condition.
Waiting for a conditon.

In addition, traps may be used to support synchronous I/O (DoIO) or asynchronous I/O.

We developed a simple kernel to support monitors on single CPU systems. The simplicity of this kernel is largely due to the technique of disabling interrupts to enforce mutual exclusion inside monitors. This technique implies that at most one monitor at a time can be active.

We also developed a multiple CPU kernel that allows more than one monitor to be active at once. This kernel is based on a virtual processor that prevents multiple simultaneous activations of the kernel.

CHAPTER 10 BIBLIOGRAPHY

The single CPU kernel presented in this chapter was developed from work on the SUE/11 operating system [Greenblatt and Holt] and from an unpublished design by C.A.R. Hoare. The multiple CPU kernel was developed from work on the SUE/360 operating system. Barnard describes the implementation of a multiple microprocessor kernel based on an earlier version of the design given in this chapter. Wirth describes another kernel for the PDP-11 that supports monitors and processes, without time slicing. Wirth's kernel supports Modula, a concurrent dialect of Pascal.

Barnard, D.T., Kulick, J.H., MacMillan, D. Hardware support for a multiprocessing kernel. Proceedings of Seventh International Symposium on Computer Systems, IEEE, Anahiem (January 1979), 47-52.

Greenblatt, I.E. and Holt, R.C. The SUE/11 operating system. *INFOR*, Canadian Journal of Operational Research and Information Processing 14,3 (October 1976), 227-232.

Wirth, N. Modula: a language for modular programming (and two other articles by Wirth in the same issue). *Software Practice and Experience Vol.* 7,1 (January-February 1977), 3-35.

CHAPTER 10 EXERCISES

1. The kernel designs presented in this chapter have no provision for error checking. For example, if a process passes the number of a nonexistent condition to the wait routine, the kernel may crash. What error checking is appropriate for a kernel? Estimate the increase in CPU time required in primitives (e.g., in wait) if error checking is incorporated. What error checking can be done at compile time to avoid run time checking by the kernel?

2. It has been suggested that the signal operation should imply an immediate return of the signaling process from the monitor. This is the case in the Concurrent Pascal language. Modify the single and multiple CPU designs for kernels to support this interpretation of signals. What effect does this have on monitor entries that return values (i.e., entries that behave like function procedures).

3. Augment the PDP-11 kernel given in this chapter to support multiple process priority conditions.

4. Implement a kernel for the System/360 (or some other computer) by translating the design of the single CPU kernel to assembly language.

5. Use a PDP-11 processor handbook and determine the time to execute wait and signal as implemented for the PDP-11 in this chapter. You will need to make certain assumptions, e.g., model of PDP-11 and average number of executions of the INSERT loop.

6. C.A.R. Hoare suggests that when a process wakes up another process using signal, the signaler should be placed on a special "urgent" queue, waiting to get back into the monitor. Each time a process exits the monitor, the urgent queue is checked and, if non-empty, a waiting signaler is given control of the monitor. The implementation of monitors for a single CPU system given in this chapter does not use an urgent queue. Instead, the signaler is put on the ready queue. Give the design of a kernel that uses an urgent queue. What logical difference (if any) does the urgent queue make to processes? What difference is there in the kernel's size and speed? What difference is there in terms of the performance of processes?

7. Throughout this chapter, we have made no mention of changing the priorities of processes (used in ordering the ready queue). In many single systems, fixed priorities may be sufficient, because the relative importance to performance of each process may be known at system construction time. However, changing the priorities dynamically can be important. Specify a primitive that allows a process to change its priority. The kernel can improve performance by giving priority to I/O bound jobs. This is equivalent to decreasing the priority of processes that use a lot of CPU time. Since the kernel can observe which jobs use most of the CPU time, it can decrease their priorities. Give a simple change to the single CPU kernel's design that automatically adjusts priorities, so I/O bound jobs are favored.

8. In the design of the multiple CPU kernel it was assumed that I/O interrupts favor those CPUs that are running low priority processes. Why was this assumption made and what would happen if it were not true?

9. What would be the result of using the following versions of enter and exit kernel:

```
EnterKernel:
    Save status of running
    loop
        TestAndSet(kernelOccupied, wasOccupied)
        exit when not wasOccupied
    end loop
    Disable interrupts
```

```
ExitKernel:
    needCPU :=
        (first ready priority < lowest running priority)
    Enable interrupts
    if needCPU then
        PreemptCPU
    end if
    kernelOccupied := false
    Restore status of running
```

10. The first kernel design (for single CPU systems) allows only one monitor to be active at a given time. The second kernel design allows multiple monitors to be active concurrently. What are the implications for operating system organization of the restriction of a single active monitor at a time?

11. The design of the single CPU kernel implements mutual exclusion in monitors by disabling and enabling interrupts. What implication does this implementation technique have for the use of monitors in organizing operating systems? (Note that the design allows only one monitor at a time to be active.)

12. A virtual processor was defined for use by the multiple CPU system. That processor gives the kernel mechanisms that are similar to the mechanisms provided by a monitor, namely, mutual exclusion and the ability to block/wakeup calling processes. Specify as a language feature a modified version of a monitor that is like this virtual processor. Give the implementation of these modified monitors in terms of the kernel's virtual processor. Note that this is equivalent to the (recursive) implementation of virtual processors in terms of a virtual processor.

13. Use the virtual processor defined for the multiple CPU system to implement semaphores.

14. Suppose your operating system supports semaphores. Show how these semaphores can be used to support monitors, without the need of a kernel (other than the one that supports semaphores).

15. Specify the functional characteristics for a multiple CPU system that supports (in the hardware) the virtual processor needed for the multiple CPU kernel.

16. Implement Signal/WaitDevice for the PDP-11 (or other computer).

17. Implement a kernel for a single CPU PDP-11 system that is based on the multiple CPU kernel given in this chapter. Your kernel will allow several monitors to be active at once. Compare the speed of your kernel with the single CPU kernel given in this chapter.

18. The single CPU kernel for the PDP-11 that was given in this chapter was made smaller and faster by the following assumption. When a process executes a trap instruction it is not currently using registers R1 to R5. Consequently, some saving and restoring of registers was avoided. What are the security implications of sometimes avoiding this saving and restoring? If your process works for (or against) the KGB and another process works against (or for) the KGB, how might your process try to spy on the other process, given the knowledge that saving/restoring registers is sometimes avoided?

Appendix

SPECIFICATION OF CONCURRENT EUCLID

By J.R. Cordy and R.C. Holt

CONTENTS OF APPENDIX

This report defines the programmming language Concurrent Euclid, or CE. CE is designed for implementing software, and is particularly suited to implementing operating systems, compilers and specialized microprocessor applications. It can also serve as the basis for implementing software which is to be formally verified.

CE consists of a subset of the Euclid programming language called Sequential Euclid or SE and a set of concurrency extensions to Euclid based on monitors. The first section of this document defines the SE language independently of Euclid. The second section describes the concurrency features added to form CE. The last section describes CE features that support separate compilation of procedures, functions, modules and monitors. Attached is material describing input/output and implementation considerations.

THE SE LANGUAGE

This section describes the SE subset of Euclid. SE is defined independently of Euclid and no previous knowledge of the Euclid programming language is required. An understanding of the basic concepts of the Pascal family of programming languages is assumed.

IDENTIFIERS AND LITERALS

An *identifier* consists of any string of at most 50 letters, digits and underscores (_) beginning with a letter. Upper and lower case letters are considered identical in identifiers and keywords, hence aa, aA, Aa and AA all represent the same identifier. Keywords and predefined identifiers of Euclid, SE and CE must not be redeclared. A list of these is given in "Keywords and Predefined Identifiers".

A *string literal* is any sequence of one or more characters not including a quote (') surrounded by quotes. Within strings, the characters quote, dollar sign, new line and end of file are represented as $', $$, $N and $E respectively. As well, $T, $S and $F may be used for tab, space, and form feed respectively.

A *character literal* is a dollar sign ($) followed by any single character. The character literals corresponding to quote, dollar sign, space, tab, form feed, new line and end of file are $$', $$$, $$S, $$T, $$F, $$N and $$E respectively.

In every implementation, the character set for string and character literals will contain at least the upper and lower case letters A-Z and a-z, the digits 0-9 and the special characters ".,:;!?()[]{}+-*/<=>'$#^&%", space, tab, form feed, new line and end of file. Character values are ordered such that A<B<C<...<Z, a<b<c<...<z and 0<1<2<...<9. Ordering of character values is implementation dependent otherwise.

An *integer literal* is a decimal number, an octal number or a hexadecimal number. A decimal number is any sequence of decimal digits. An octal number is any sequence of octal digits followed by #8. A hexadecimal number is any sequence of hexadecimal digits (represented as the decimal digits plus the capital letters A through F) beginning with a decimal digit and followed by #16. Negative values are obtained using the unary - operator; see "Expressions".

SOURCE PROGRAM FORMAT

A *comment* is any sequence of characters not including comment brackets surrounded by the comment brackets { and }. Comments may cross line boundaries.

A *separator* is a comment, blank, tab, form feed or source line boundary. Programs are free-format; that is, the identifiers, keywords, literals, operators and special characters which make up a program may have any number of separators between them. Separators cannot be embedded in identifiers, keywords, literals or operators, except that blanks may appear as part of the value of a string literal. Identifiers, keywords and literals must not cross line boundaries. At least one separator must appear between adjacent identifiers, keywords and literals.

SYNTACTIC NOTATION

The following sections define the syntax of SE.

The following notation is used:

> {item} means zero or more of the item
> [item] means the item is optional

Keywords and special symbols are given in **bold face**.

The following abbreviations are used:

> id for identifier
> expn for expression
> typeDefn for typeDefinition

Semicolons are not required, but they may optionally appear following statements, declarations and import, export and checked clauses.

PROGRAMS

A main program consists of a module declaration.

A *program* is:

moduleDeclaration

Execution of a program consists of initializing the main module, see "Modules".

Modules, procedures and functions can be compiled separately; see "Separate Compilation".

MODULES

A *moduleDeclaration* is:

> **var** id :
> > **module**
> > > **[imports ([var]** id {, **[var]** id})]**
> > > **[exports (** id {, id})]**
> > > [**[not] checked**]
> > > {declaration In Module}
> > > [**initially**
> > > > procedureBody]
> > **end module**

Execution of a module declaration consists of executing the declarations in the module and then the **initially** procedure of the module. Execution of a program consists of executing the main module declaration in this way.

Module declarations may be nested inside other modules but must not be nested inside procedures and functions.

A module defines a package of variables, constants, types, procedures and functions. The interface of the module to the rest of the program is defined by its **imports** and **exports** clauses.

The **imports** clause lists the global identifiers which are to be visible inside the module. Variable, collection and module identifiers may be imported **var** (or not). Imported variables can be assigned to or passed as **var** parameters within the module only if they are imported **var**. Elements of an imported collection can be allocated, freed, assigned to or passed as **var** parameters only if the collection is imported **var**. Procedures of an imported module can be called only if the module is imported **var**. Imported identifiers must not be redeclared inside the module.

The **exports** clause lists those identifiers defined inside the module which may be accessed outside the module using the **.** operator. Exported variables cannot be assigned to or passed as **var** parameters outside the module. Elements of exported collections cannot be allocated, freed, assigned to or passed as **var** parameters outside the module. Unexported identifiers cannot be referenced outside the module.

Named types declared inside a module are *opaque* outside the module, that is, they are not considered equivalent to any other type. Variables and constants whose type is opaque cannot be subscripted, field selected or compared.

Modules may be **checked**; this causes all **assert** statements, subscripts and **case** statements in the module to be checked for validity at run-time. In addition, a particular implementation may check other conditions such as ranges in assignments and overflow in expressions. Modules not already nested inside an unchecked module are checked by default and must be explicitly declared **not checked** to turn off run-time checking.

Even though declared like variables, modules are not variables and cannot be assigned, compared, passed as parameters or exported.

Modules can be separately compiled if desired; see "Separate Compilation".

DECLARATIONS

A *declarationInModule* is one of the following:

- a. constantDeclaration
- b. variableDeclaration
- c. typeDeclaration
- d. variableBinding
- e. moduleDeclaration
- f. collectionDeclaration
- g. procedureDeclaration
- h. functionDeclaration
- i. converterDeclaration
- j. **assert** [(expn)]

Forms (a) through (i) are declarations for new identifiers as explained in the following sections. Form (j) is an **assert** statement; see "Statements". An identifier must be declared textually preceding any references to it.

CONSTANT DECLARATIONS

A *constantDeclaration* is one of:

    a.   [**pervasive**] **const** id := manifestExpn
    b.   [**pervasive**] **const** id : typeDefn := expn
    c.   [**pervasive**] **const** id : typeDefn :=
        (manifestExpn {, manifestExpn})
    d.   [**pervasive**] **const** id := stringLiteral

A constantDeclaration gives a name to a value which is constant throughout the scope of the declaration. The value of a scalar constant can be *manifest* or *nonmanifest*. A manifest expression is one whose value is known at compile-time (see "Manifest Expressions"). A nonmanifest expression must be evaluated at run-time. Non-scalar values are always considered nonmanifest.

Form (a) defines a *manifest named constant*. The type of the constant is the type of the value expression, which must be manifest. Manifest named constants are not represented at run time since their values are always known at compile time.

Form (b) declares a *nonmanifest named constant* of the specified type. The value of the expression may be manifest or nonmanifest, and must be assignable to the constant's type. References to nonmanifest named constants are always considered nonmanifest even if their value is manifest.

Form (c) declares an array constant. The typeDefn must be an array type or named array type whose component type is scalar. The list of expressions gives the values of the elements of the array constant. The element values must be manifest expressions assignable to the element type of the array. The number of element values specified must be exactly the number of elements in the array.

Form (d) allows declaration of an array constant using a string literal value. The type of the constant is **packed array** 1..n of Char where n is the length of the string literal.

Constants declared using **pervasive** are automatically imported into all subscopes of the scope in which they are declared. Such constants need not be explicitly imported.

VARIABLE DECLARATIONS

A *variableDeclaration* is:

    [**register**] **var** id [(**at** manifestExpn)] : typeDefn
        [:= expn]

A variableDeclaration declares a variable of the specified type. The **at** clause declares a variable at an absolute machine location. Variables may optionally be declared with an initial value which is assigned to the variable when the declaration is executed. The initial value expression must be assignable to the variable's type.

Local variables in procedures and functions may optionally be declared **register**. This is a hint to the compiler that it should attempt to allocate the variable to a register. Register variables cannot be bound to nor passed to a reference parameter. A **register** variable declaration cannot have an **at** clause.

TYPES AND TYPE DECLARATIONS

A *typeDeclaration* is:

> [**pervasive**] **type** id = typeBody

The *typeBody* is one of:

> a. typeDefn
> b. **forward**

A typeDeclaration gives a name to a type. The type name can subsequently be used in place of the full type definition. A named type is equivalent to the type that it names (except when exported, see "Type Equivalence and Assignability").

Named types may optionally be declared **pervasive**. Type names declared using **pervasive** are automatically imported into all subscopes of the scope in which they are declared. Such types need not be explicitly imported.

Form (b) declares a forward type. A forward type declares a type name whose definition will be given in a later type declaration in the scope. A forward type can be used only as the element type of a **collection** until its real type definition is given. This allows the declaration of collections whose elements contain pointers to other elements in the collection.

A *typeDefn* is one of the following:

> a. standardType
> b. manifestConstant **..** manifestExpn
> c. [**packed**] **array** indexType **of** typeDefn
> d. **set** of baseType
> e. [**packed**] recordType
> f. pointerType
> g. namedType

The *standardTypes* are:

SignedInt   - signed integer, implementation
                                defined range (at least -32768..32767)

UnsignedInt   - unsigned integer, implementation
                                defined range (at least 0..65535)

LongInt   - signed integer, implementation
                                defined range (typically 32 bits)

ShortInt   - unsigned integer, implementation
                                defined range (typically a byte)

Boolean   - values are "true" and "false"

Char   - single character

StorageUnit   - no operations or literals, smallest
                                addressable memory unit (typically a byte)

AddressType   - implementation defined integer range

The standard types and the constants true and false are implicitly declared pervasive in the global scope and need not be imported.

Form (b) is a *subrange* type. The leading constant must be a (possibly negated) literal or manifest named constant and gives the lower bound of the range of values of the type. The expression, which must be manifest, gives the upper bound of the range. The bounds must be both integer values or both character values. The lower bound must be less than or equal to the upper bound.

A *scalar* type is a subrange, pointer or one of the standard types.

Form (c) is an array type. The *indexType* must be a subrange type, Char or a named type which is an indexType. The indexType gives the range of subscripts. The typeDefn gives the type of the elements of the **array**.

Elements of an array variable are referenced using subscripts (see "Variables and Constants") and themselves used as variables. Array variables and constants may be assigned (but not compared) as a whole.

Arrays can be **packed**, which allows the compiler to pack the elements more efficiently. The type of string literals is "**packed array** 1..n of Char" where n is the length of the string.

Form (d) is a **set** type. The *baseType* of the set must be a subrange of integer with lower bound 0 or a namedType which is a baseType. An implementation may limit the upper bound of a set type to insure efficient code; this limit will be at least 15.

A *recordType* is:

```
record
    var id : typeDefn
    {var id : typeDefn}
end record
```

Variables declared using a **record** type have the fields given by the variable declarations in the recordType. Fields of a record variable may be referenced using the . operator (see "Variables and Constants") and themselves used as variables. Record variables may be assigned (but not compared) as a whole.

The variable declarations in a record type must not have initial values and cannot be declared using **register** or **at** clauses.

Records can be **packed**, which allows the compiler to pack the elements more efficiently.

A *pointerType* is:

    ^ collection Id

Variables declared using a pointerType are pointers to dynamically allocated and freed elements of the specified collection; see "Collections". Pointer variables are used as subscripts of the specified collection to select the element to which they point. The selected element can be used as a variable. Pointer variables may be assigned, compared for equality and passed as parameters.

A *namedType* is:

    [moduleId .] typeId

The typeId must be a previously declared type name. Type names exported from a module are referenced outside the module using the . operator.

TYPE EQUIVALENCE AND ASSIGNABILITY

Two types are defined to be *equivalent* if they are

(a) subranges with equal first and last values
(b) arrays (both packed or both unpacked) with
    equivalent index types and equivalent component types
(c) sets with equivalent base types
(d) pointers to the same collection

A declared type identifier is equivalent to the type it names, with the following exception. When an exported type identifier is used outside its module, as "moduleId.typeId", it is a unique type, equivalent to no other type.

Each type definition for a record type creates a new type that is not equivalent to any other record type definition.

An array value can be assigned to an array variable, a record value assigned to a record variable, a set value assigned to a set variable and a pointer value assigned to a pointer variable only if the source and target of the assignment have equivalent types.

An expression can be assigned to a scalar variable only if (i) the "root" type of the expression and the "root" type of the variable are equivalent, and (ii) the value of the expression is in the range of the variable's type. The "root" type of Char and character subrange types is Char. The root type of SignedInt, UnsignedInt, LongInt, ShortInt, AddressType and integer subranges is integer. The root type of any other type is the type itself.

A variable can be passed to a reference parameter only if its type is equivalent to the parameter type. An expression can be passed to a value parameter only if it is assignable to the parameter type; see "Procedures and Functions".

VARIABLE BINDINGS

A *variableBinding* is one of:

    a. **bind** [**register**] [**var**] id **to** variable
    b. **bind** ([**register**] [**var**] id **to** variable
             {, [**register**] [**var**] id **to** variable})

A variableBinding declares a new identifier for an arbitrary variable reference which may contain subscripts and . operators. The new identifier is subsequently used in place of the variable reference within the scope in which the binding appears. If the bound variable is to be assigned to or passed to a var parameter, the binding must be declared using "var". SE does not allow "aliasing" of variables (i.e., having two names for the same variable in a scope). Hence the "root" variable (the first identifier in the variable reference) becomes inaccessible for the scope of the binding.

Form (b) allows bindings to different elements or fields of the same variable or module. Since SE does not allow aliasing of variables, bindings to the same field, element or variable are not allowed.

Local binds in procedures and functions may optionally be declared **register**. This is a hint to the compiler to attempt to allocate the bind to a register.

Elements of packed arrays and fields of packed records cannot be bound to.

COLLECTIONS

A *collectionDeclaration* is:

> **var** id : **collection of** typeDefn

A collection is essentially an array whose elements are dynamically allocated and freed at run-time. Elements of a collection are referenced by subscripting the collection name with a variable of the collection's pointer type. This subscripting selects the particular element of the collection located by the pointer variable.

Elements of a collection are allocated and freed dynamically by calls to the built-in operations New and Free. "C.New(p)" allocates a new element in the collection C and sets p to point at it. If no more space is available then p is set to "C.nil". "C.Free(p)" frees the element of C pointed at by p and sets p to "C.nil". In each case p is passed as a **var** parameter and must be a variable of the pointer type of C. These operations are invoked as statements in procedures, see "Statements". They cannot be used in functions.

The built-in constant "C.nil" is the null pointer value for the collection.

Collections themselves cannot be assigned, compared or passed as parameters.

PROCEDURES AND FUNCTIONS

A *procedureDeclaration* is:

> **procedure** id [([**var**] id : parameterType
> {, [**var**] id : parameterType})] =
> procedureBody

A *functionDeclaration* is:

> **function** id [(id : parameterType
> {, id : parameterType})]
> **returns** id : resultType =
> procedureBody

A procedure is invoked by a procedure call statement, with actual parameters if required. A function is invoked by using its name, with actual parameters if required, in an expression.

A procedure may return explicitly by executing a return statement or implicitly by reaching the end of the procedure body. A function must return via "return(expn)".

Procedures and functions may optionally take parameters, the types of which are defined in the header. The parameters can be referred to inside the procedure or function using the names declared in the header. Parameters to a procedure may be declared using **var**, which means the parameter may be assigned to or further passed as a **var** parameter inside the procedure. Parameters declared without using **var** are constants and cannot be assigned to or passed as **var** parameters. Functions are not allowed to have any side-effects and cannot have **var** parameters. Only variable references can be passed to **var** parameters.

A parameter is a *reference* parameter if it is declared using "var" or if its type is an array or record. Other parameters are *value* parameters. Hence, a value parameter is a non-var parameter whose type is a scalar or set.

A *parameterType* is one of:

  a. typeDefn
  b. [**packed**] **array** manifestConstant **.. parameter of**
      typeDefn
  c. **universal**

The type of a variable, record or array passed to a reference parameter must be equivalent to the parameter's type with the following exceptions. (1) The upper bound of the index type of an array parameter can be declared using the keyword **parameter** in which case any array whose element type and index type lower bound are equivalent to the parameter's can be passed to the parameter. (2) The type of a parameter can be specified as **universal**, in which case a variable or non-manifest named constant of any type can be passed to the parameter. Inside the procedure, a **universal** parameter is equivalent to a parameter of type array 1..parameter of StorageUnit, where the upper bound is the size of the actual parameter in StorageUnits. Parameters declared using parameter or **universal** do not have the ".size" standard component and cannot be assigned or compared as a whole. (Note: Full Euclid does not allow forms (b) and (c).)

The type of an expression passed to a value parameter must be assignable to the parameter's type.

SE does not allow "aliasing" of variables (i.e., having two names for a given variable or part of a given variable in the same scope). Hence a variable or part of a variable which is imported directly or indirectly into a procedure cannot be passed to a reference parameter of the procedure. (A variable is directly imported if it appears in the procedure's import list. It is indirectly imported if an imported module or procedure directly or indirectly imports it.)

Elements of packed arrays and fields of packed records cannot be passed to reference parameters.

The **returns** clause defines the result type of a function. The return identifier is required for compatibility with full Euclid but cannot be referenced.

A *resultType* is one of:

   a.   standardType
   b.   manifestConstant .. manifestExpn
   c.   **set of** baseType
   d.   pointerType
   e.   namedType

The result type of a function must be a scalar type or set. The expression in a function's **return** statement must be assignable to the result type.

A *procedureBody* is:

```
[imports ( [var] id {, [var] id} )]
begin
   [[not] checked]
   {declarationInRoutine}
   {statement}
end [id]
```

The identifier following the **end** must be the procedure or function identifier. If the procedure is the **initially** procedure of a module then the **end** identifier must not be present.

The imports clause of a procedure or function specifies those global identifiers which are to be visible inside the procedure or function. Only those variables imported into a procedure using **var** may be assigned to or passed to a **var** parameter inside the procedure. Functions are not allowed to have side-effects and cannot import anything **var**. This restriction is transitive; hence a function cannot import a procedure which imports anything **var**. A procedure or function which is recursive must explicitly import itself.

Procedures and functions may be **checked**; this causes **assert** statements, subscripts and **case** statements to be checked for validity at runtime. In addition, a particular implementation may check other conditions, such as ranges in assignments and overflow in expressions. Procedures and functions not nested inside an unchecked module are checked by default and must be explicitly declared **not checked** to turn off run-time checking.

A procedure returns when it executes a return statement or reaches the end of the procedure. A function is executed similarly but must return via **return**(expn).

Procedures and functions can be separately compiled; see "Separate Compilation".

A *declarationInRoutine* is one of:

   a.   constantDeclaration

    b.   variableDeclaration

    c.   typeDeclaration

    d.   variableBinding

    e.   collectionDeclaration

    f.   converterDeclaration

    g.   **assert** [(expn)]

Modules, procedures and functions cannot be nested inside a procedure or function. Form (g) allows **assert** statements to appear in declaration lists.

TYPE CONVERTERS

A *converterDeclaration* is:

    converter id (typeId) **returns** typeId

A converterDeclaration declares a type converter. A type converter can be used to convert a variable or nonmanifest named constant to a type other than its declared type. Both the parameter and result type of a type converter must be named or standard types. An implementation is not expected to generate any code for a type conversion.

The type of a converted variable or constant must be equivalent to the converter's parameter type. Expressions, literals, manifest values, elements of packed arrays and fields of packed records cannot be type converted.

If the size of the target type is larger than the size of the source type, or the alignment of the target type is more constrained than the alignment of the source type, then the conversion may be meaningless.

STATEMENTS

A *statement* is one of:

    a.   variable := expn

    b.   [moduleId.] procedureId [(expn {, expn})]

    c.   **assert** [(expn)]

    d.   **return** [(expn)]

    e.   **if** expn **then**
           {statement}
       {**elseif** expn **then**
           {statement}}
       [**else**
           {statement}]
       end if

    f.   **loop**
           {statement}
       end loop

g.   **exit** [**when** expn]
h.   **case** expn **of**
           manifestExpn {, manifestExpn} = >
               {statement}
               end manifestExpn
           {manifestExpn {, manifestExpn} = >
               {statement}
               end manifestExpn}
           [**otherwise** = >
               {statement}]
       end case
i.   **begin**
           {declarationInRoutine}
           {statement}
       end
j.   collectionId . New (variable)
k.   collectionId . Free (variable)

Form (a) is an assignment statement. The expression is evaluated and the value assigned to the variable. The expression must be assignable to the variable type; see "Type Equivalence and Assignability".

Form (b) is a procedure call. An exported procedure is called outside the module in which it was declared using the . operator.

The type of an expression passed to a value parameter must be assignable to the parameter's type. The type of a variable or value passed to a reference parameter must be equivalent to the parameter's type. If the upper bound of the type of an array parameter is declared using **parameter**, any array whose element type and index type lower bound are equivalent to the parameter's can be passed to the parameter.

An actual parameter passed to a **var** parameter must be a variable, a bound variable or a **var** formal parameter. If it is an imported variable, it must have been imported using **var**. Since SE does not allow aliasing of variables, a variable or part of a variable which is passed to a reference parameter cannot be passed to another reference parameter of the same call.

Form (c) is an **assert** statement. The parenthesized expression is optional; if it is omitted, it can be replaced by a comment. If present, it must be of type Boolean. The expression is evaluated and checked at run time if it appears in a checked scope. Assert statements may appear in both statement lists and declaration lists. They cannot appear inside records.

Form (d) is a **return** statement. The **return** statement causes an immediate return from the procedure or function when executed. The optional parenthesized expression gives the value to be returned from a function. The return expression must be assignable to the function's result type. The return expression is required for function returns. It is forbidden for procedure returns. A function must return via a return statement and not implicitly by reaching the end of the function body. A procedure may return either via a return statement or implicitly by reaching the end of the procedure body.

Form (e) is an **if** statement. The conditional expression following **if** and each **elseif** is evaluated until one of them is found to be true, in which case the statements following the corresponding **then** are executed. If none of the expressions evaluates to true then the statements following **else** are executed; if no **else** is present then execution continues following the **if** statement. The conditional expressions must be of type Boolean.

Form (f) is the looping construct. The statements within the loop are repeated until one of its **exit** statements or a **return** statement is executed.

Form (g) is a loop **exit**. When executed, it causes an immediate exit from the nearest enclosing loop. The optional **when** expression makes the exit conditional. If the expression, which must be Boolean, evaluates to true then the **exit** is executed, otherwise execution of the loop continues. An **exit** statement cannot appear outside a loop.

Form (h) is a **case** statement. The **case** expression is evaluated and used to select one of the alternative labels. The statements which follow the matching label value are executed. If the **case** expression value does not match any of the label values then the statements following **otherwise** are executed. If no **otherwise** is present, the case expression must match one of the label values. When execution of the statements following the selected label is completed, execution continues following the case statement.

The root type of the **case** expression must be integer or Char. All of the label expressions must have the same root type as the **case** expression. Label expressions must be manifest, i.e., their values must be known at compile time. The values of all label expressions in a given **case** statement must be distinct. The value of the manifest expression following the **end** of an alternative must be equal to the first label expression of the alternative.

An implementation may limit the range of **case** label expression values to insure efficient code; this range will include at least the ranges of Char and ShortInt.

Form (i) is a **begin** block. Begin blocks can be used to group local declarations within a procedure or function. In particular, they can be used to make local binds.

Forms (j) and (k) are the built-in collection operations New and Free (see "Collections").

VARIABLES AND CONSTANTS

A *variable* is:

> [moduleId .] id {componentSelector}

The syntax for variables includes variable and constant references. An exported variable or constant is referenced outside the module in which it is declared using the . operator.

A *componentSelector* is one of:

> a. (expn)
> b. . id

Form (a) allows subscripting of variable and constant arrays. The type of the subscript expression must be assignable to the index type of the array. The value of the subscript expression must be in the declared range of the index type of the array. Subscripts which appear in checked scopes are checked for validity at run-time.

Form (a) also allows references to elements of a collection. In this case, the subscript expression must be a pointer to an element of the collection.

Form (b) allows record field selection. Fields of a record variable are referenced using the . operator.

Form (b) also allows standard component references (see "Standard Components").

EXPRESSIONS

An *expn* is one of the following:

> a. variable
> b. literalConstant
> c. setTypeId (elementList)
> d. collectionId . nil
> e. [moduleId .] functionId [(expn {, expn})]
> f. [moduleId .] converterId (expn)
> g. (expn)
> h. - expn
> i. expn arithmeticOperator expn
> j. expn comparisonOperator expn
> k. **not** expn
> l. expn booleanOperator expn
> m. expn setOperator expn

The arithmeticOperators are +, -, * (multiply), **div** (truncating integer divide) and **mod** (integer remainder). The **mod** operator is defined by "x **mod** y = x - y*(x **div** y)". Operands of the arithmetic operators and unary minus must be integers or expressions having root type integer. The arithmetic operators yield an integer result. (Note: +, - and * are also **set** operators; see below.)

The comparisonOperators are <, >, =, <=, >= and "not =". Operands of comparison operators must either have equivalent types or the same root type; see "Type Equivalence and Assignability". The comparison operators yield a Boolean result. Arrays and records cannot be compared. Sets and Boolean expressions can be compared for equality only. (Note: <= and >= are also **set** operators; see below.)

The booleanOperators are **and** (intersection), **or** (union) and -> (implication). The Boolean operators and the **not** operator take Boolean operands and yield a Boolean result. The Boolean operators are conditional; that is, if the result of the operation can be determined from the value of the first operand then the second operand is not evaluated.

The **set** operators are + (set union), - (set difference), * (set intersection), <= and >= (**set** inclusion), and **in** and **not in** (element containment). The **set** operators +, - and * take operands of equivalent set types and yield a **set** result. The **set** operators <= and >= take operands of equivalent set types and yield a Boolean result. The operators **in** and **not in** take a set as right operand and an integer expression as left operand. They yield a Boolean result.

The order of precedence is among the following classes of operators (most binding first):

1. unary -
2. *, **div**, **mod**
3. +, -
4. <, >, =, <=, >=, **not** =, **in**, **not in**
5. **not**
6. **and**
7. **or**
8. ->

Expression form (a) includes references to constants and variables including elements of arrays and collections, fields of records, and constants and variables exported from a module.

Form (b) includes integer, character and string literal constants.

Form (c) is a **set** constructor. The setTypeId must be the name of a **set** type. The **set** constructor returns a set containing the specified elements.

An *elementList* is one of:

    a. [expn {, expn}]
    b. **all**

The element list is a (possibly empty) list of expressions of the base type of the set, or **all**. If **all** is specified, the constructor returns the complete set. If no elements are specified, the constructor returns the empty set.

Expression form (d) is the null pointer value of the specified collection.

Form (e) is a function call. Functions exported from a module are referenced outside the module using the . operator. An actual parameter to a function must be an expression assignable to the parameter type.

Form (f) is a type conversion. The type of the actual parameter is changed to the result type of the type converter. The actual parameter must be a variable or nonmanifest named constant whose type is equivalent to the source type of the converter. Type converters exported from a module are referenced outside the module using the . operator.

BUILT-IN FUNCTIONS

SE has three built-in functions, Chr, Ord and Long. "Chr(i)" returns the character whose machine representation is the positive integer value i. "Ord(c)" returns the positive integer machine representation of the character c. Chr and Ord are defined such that for all characters "c" in the machine character set, Chr(Ord(c)) = c. "Long(i)" forces the integer expression i to be extended to LongInt precision; see "Precision of Arithmetic". (Note: In full Euclid, the Ord built-in function is called "Char.Ord".)

STANDARD COMPONENTS

SE defines two *standard components*, size and address. "T.size" returns the length in StorageUnits (typically bytes) of the machine representation of the variable or type T. "V.address" returns the AddressType machine address of the variable V. The size and address standard components are not allowed for elements of packed arrays and fields of packed records. The address standard component is not allowed for variables declared **register**.

MANIFEST EXPRESSIONS

A manifest expression is an expression whose value can be computed as a literal constant at compile time. The extent of such compile-time computation is implementation dependent, but every implementation will consider at least the following to be manifest:

1. Integer and Char literal constants
2. The Boolean values "true" and "false"
3. Manifest named constants
4. The arithmetic operations unary -, +, -, *, **div** and **mod** when both operands are manifest and both the operands and result lie in the range of SignedInt (at least -32768..32767)
5. The built-in functions Chr and Ord when the actual parameter is manifest

A *manifestExpn* is an expression whose value is manifest. A *manifestConstant* is a (possibly negated) literal constant or manifest named constant.

PRECISION OF ARITHMETIC

The precision of an arithmetic operation or comparison is determined by the precision of the operands. Operands have one of three precisions which correspond to the standard types SignedInt, UnsignedInt and LongInt.

The precision of a variable or non-manifest named constant operand is determined by its declared type. If its type is SignedInt, ShortInt or any subrange whose bounds both lie in the range of SignedInt then its operand precision is SignedInt. If its type is UnsignedInt or any subrange whose bounds both lie in the range of UnsignedInt but not in SignedInt then its precision is UnsignedInt. Otherwise, its precision is LongInt.

The precision of a literal or manifest named constant operand is SignedInt if its value lies in the range of SignedInt, UnsignedInt if its value lies in the range of UnsignedInt but not of SignedInt, and LongInt otherwise.

The precision of an arithmetic operation or comparison is LongInt if at least one operand has LongInt precision, UnsignedInt if at least one operand has UnsignedInt precision and neither has LongInt precision, and SignedInt otherwise.

The precision of the result of an arithmetic operation is the precision of the operation. Every implementation will guarantee to obtain the arithmetically correct result if the result of an operation lies within the range of the result precision. If the arithmetically correct result lies outside the range of the result precision then the result may be meaningless.

Note that the precision of an operation or comparison can always be forced to LongInt by extending the precision of one or both of the operands using the Long built-in function (see "Built-in Functions").

SOURCE INCLUSION FACILITY

Other source files may be included as part of a program using the **include** statement.

An *includeStatement* is:

> **include** stringLiteral

The stringLiteral gives the name of a source file to be included in the compilation. The **include** statement is replaced in the program source by the contents of the specified file.

Include statements can appear anywhere in a program and can contain any valid source fragment. Included source files can themselves contain **include** statements.

CONCURRENCY FEATURES

The Concurrent Euclid (CE) language is an extension of SE designed to allow concurrent programming with monitors. SE is a subset of Euclid but CE is not, because concurrency and monitors are not features of Euclid.

The concurrency features of CE will be presented in the following order:

(1) processes, reentrant procedures and modules;
(2) monitors, entry procedures and functions;
(3) conditions, signalling and waiting;
(4) simulation and the **busy** statement.

PROCESSES

Each CE module (including the main module) can have any number of concurrent processes in it.

A *moduleDeclaration* is:

> **var** id :
> module
>   **[imports (** [**var**] id {, [**var**] id} **)]**

```
        [exports ( id {, id} )]
        [[not] checked]
        {declaration In Module}
        [initially
            procedureBody]
        {process id [( memoryRequirement )]
            procedureBody}
    end module
```

Each **process** is like a parameterless **procedure**. Concurrent execution of the processes of the module begins following execution of the **initially procedure** of the module. A **process** terminates by executing its last statement or by executing a **return** statement in its body. The **process** identifier is for documentation only since processes cannot be called.

Processes can communicate with each other by changing and inspecting variables declared in the module or imported into it. Generally, however, processes communicate by means of monitors.

Each process requires a certain amount of memory space for its variables. When the process calls a procedure or function, the requirement increases to provide space for the new local variables. When the procedure or function returns, the requirement decreases to its former amount. The programmer can provide his own estimate of the process's required space as a parenthesized manifest integer expression following the keyword **process**. This estimate is in StorageUnits (normally bytes) and can be based on previous program executions. If this estimate is omitted, the implementation provides a default space allocation.

All procedures and functions declared in a CE program are *reentrant*, meaning that they can be executed simultaneously by more than one process.

Modules, monitors, procedures and functions cannot be nested inside a **process**.

MONITORS

A *monitor* is essentially a special kind of module which implements inter-process communication with synchronization.

A *declaration In Module* is one of the following:

    a.   constant Declaration
    b.   variable Declaration
    c.   type Declaration
    d.   variable Binding
    e.   module Declaration
    f.   monitor Declaration

g.   collectionDeclaration
h.   procedureDeclaration
i.   functionDeclaration
j.   converterDeclaration
k.   **assert** [(expn)]

Monitors may only be declared inside modules. Monitors cannot be nested inside procedures, functions or other monitors.

A *monitorDeclaration* is:

> **var** id :
> > **monitor**
> > > [**imports** ([**var**] id {, [**var**] id})]
> > > [**exports** (id {, id})]
> > > [[**not**] **checked**]
> > > {declarationInMonitor}
> > > [**initially**
> > > > procedureBody]
> >
> > **end monitor**

The imports list of a monitor specifies the global identifiers which are accessible inside the monitor, exactly like the imports list in a module.

The exports list of a monitor specifies those identifiers defined inside the monitor which may be accessed outside the monitor using the . operator. Unlike modules, monitors cannot export variables.

Procedures and functions which are exported from a monitor are called monitor *entries*. Entry procedures and functions of a monitor cannot be invoked inside the monitor. Outside the monitor, entry procedures and functions can be invoked exactly like the procedures and functions of a module, using the . operator.

Procedures and functions which are entries of a monitor cannot be separately compiled except as part of the entire monitor.

It is guaranteed that only one process at a time will be executing inside a monitor. As a result, mutually exclusive access to a monitor's variables is implicitly provided, since a monitor cannot export any variables. If a process calls an entry of a monitor while another process is executing in the monitor, the calling process will be blocked and not allowed in the monitor until no other process is executing in the monitor.

A *declarationInMonitor* is one of the following:

a.   constantDeclaration
b.   variableDeclaration
c.   typeDeclaration
d.   variableBinding
e.   conditionDeclaration

    f.   collectionDeclaration
    g. procedureDeclaration
    h. functionDeclaration
    i.   converterDeclaration
    j.   **assert** [(expn)]

Modules and monitors cannot be declared inside a monitor. A monitor cannot contain a nested process.

Monitors can be separately compiled; see "Separate Compilation".

CONDITIONS

A *conditionDeclaration* is one of:

    a. **var** id : [**priority**] **condition**
    b. **var** id : **array** indexType **of** [**priority**] **condition**

The only place a condition can be declared is as a field of a monitor. The only allowed use of conditions is in the **wait** and **signal** statements and in the "empty" built-in function. Conditions cannot be assigned, compared or passed as parameters. Arrays of conditions are allowed. Conditions may be imported **var** (or not). An imported condition can be used in a **wait** or **signal** statement only if it is imported **var**.

Two new statements are introduced:

    wait (conditionVar [, priorityValue])
    signal (conditionVar)

Where a *conditionVar* is:

    conditionId [(expn)]

The **wait** and **signal** statements each specify a conditionVar. Each of these must be a conditionId or a subscripted condition array. These statements can appear only in monitors, but not in a monitor's **initially** procedure.

When a process executes a wait statement for condition C it is blocked and is removed from the monitor. When a process executes a **signal** statement for condition C, one of the processes (if there are any) waiting for condition C is unblocked and allowed immediately to continue executing the monitor. The signalling process is temporarily removed from the monitor and is not allowed to continue execution until no processes are in the monitor. If no processes were waiting for condition C, the only effect of the **signal** statement is that the signalling process may be removed from the monitor. The signalling process cannot in general know whether other processes have entered the monitor before the signaller continues in the monitor.

If the condition variable is declared with the **priority** option, the **wait** statement must specify a priority value; otherwise the priority value is not allowed in **wait**. The *priorityValue* is a SignedInt expression that must evaluate to a nonnegative integer value. The processes waiting for a priority condition are ranked in order of their specified priority values, and the process with the smallest priority value is the first to be unblocked by a signal statement.

In the case of processes waiting for non-priority conditions, or waiting with identical priorities for a priority condition, the scheduling is "fair", meaning that a particular waiting process will eventually be unblocked given enough signals on the condition.

A predefined function named "empty" accepts a condition as a parameter. It returns the Boolean value "true" if no processes are waiting for the condition, otherwise "false". Like **wait** and **signal**, "empty" can appear only inside a monitor, but not in the **initially** procedure of a monitor.

The variables in a monitor represent its state. For example, if a monitor allocates a single resource, only one variable inside the monitor is needed and it can be declared as Boolean. When this variable is true, it represents the state in which the resource is available, when false it represents the state of being allocated. When a process enters the monitor and finds that it does not have the desired state, the process leaves the monitor and becomes blocked by executing a **wait** statement on a condition. The condition corresponds to the state that the process is waiting for. Suppose a process enters a monitor and changes its state to a state that may be waited for by other processes. The process should execute a **signal** statement for the condition corresponding to the new state. If there are processes waiting for this state transition, then they will be blocked on the condition, and one of them will immediately resume execution in the monitor. Because of this immediate resumption, the signalled process knows the monitor is in the desired state, without testing monitor variables. The signalling process is allowed to continue executing only when no other processes are in the monitor. If no processes were waiting on the condition, the only effect of the **signal** statement is to temporarily remove the signaller from the monitor.

As specified by Hoare, monitors and conditions are intended to be used in the following manner. The programmer should associate with the monitor's variables a consistency criterion. The consistency criterion is a Boolean expression that should be true between monitor activations, or whenever a process enters or leaves a monitor. Hence, the programmer should see that it is made true before each **signal** or **wait** statement in the monitor and before each return from an entry of the monitor. The programmer should also associate a Boolean expression, call it Ei, with each condition Ci. The expression Ei should be true whenever a **signal** is executed for condition Ci. A process that is unblocked after waiting for a condition knows that Ei is true because the signalled process (not the signalling process) executes first. (The consistency criterion and each Ei for a condition do not necessarily appear as executable code in the monitor.) In general, when a process changes the monitor's state so that one of the awaited relations Ei becomes true, the corresponding condition Ci should be signalled.

THE BUSY STATEMENT

A statement is introduced to allow simulation using timing delays:

busy (time)

The *time* must be a nonnegative SignedInt expression. The **busy** statement can be understood in terms of simulated time recorded by a system clock. This clock is **set** to zero at the beginning of execution of a program. With the exception of the **busy** statement (or **wait** statements causing an indirect delay for a **busy** statement), statements take negligible simulated time to execute. When the programmer wants to specify that a certain action takes time to complete, the **busy** statement is used. The process that executes the **busy** statement is delayed until the system clock ticks (counts off) the specified number of time units.

SEPARATE COMPILATION

This section describes the extensions made to CE to allow separate compilation of procedures, functions, modules and monitors.

EXTERNAL DECLARATIONS

Procedures, functions, modules and monitors may be declared "external", which means that they are to be separately compiled and joined with the program at link time. Due to linker restrictions, a particular implementation may be forced to place a limit on the number of significant characters in external module, monitor, procedure and function identifiers.

An *externalProcedureDeclaration* is:

> **procedure** id [([**var**] id : parameterType
> {, [**var**] id : parameterType})] =
> **external**

An *externalFunctionDeclaration* is:

```
function id [( id : parameterType
             {, id : parameterType} )]
          returns id : resultType =
external
```

An *externalModuleDeclaration* is:

```
var id :
    external module
        [imports ( [var] id {, [var] id} )]
        [exports ( id {, id} )]
        {declarationInExternalModule}
    end module
```

A *declarationInExternalModule* is one of:

    a.   manifestConstantDeclaration
    b.   typeDeclaration
    c.   collectionDeclaration
    d.   converterDeclaration
    e.   externalProcedureDeclaration
    f.   externalFunctionDeclaration

An *externalMonitorDeclaration* is:

```
var id :
    external monitor
        [imports ( [var] id {, [var] id} )]
        [exports ( id {, id} )]
        {declarationInExternalMonitor}
    end monitor
```

A *declarationInExternalMonitor* is one of:

    a.   manifestConstantDeclaration
    b.   typeDeclaration
    c.   collectionDeclaration
    d.   converterDeclaration
    e.   externalProcedureDeclaration
    f.   externalFunctionDeclaration

An external declaration can appear in place of the real declaration and specifies that the corresponding procedure, function, module or monitor is to be compiled separately.

Processes and **initially** procedures of modules cannot be declared external. Procedures and functions which are entries of a monitor cannot be declared external except as part of an external monitor declaration. Nonmanifest and array named constants cannot be declared in an external module or monitor.

COMPILATIONS

A compilation can consist of a main program (see "Programs") or a separate compilation.

A *separateCompilation* is:

{separateDeclaration}

Each *separateDeclaration* is one of the following:

    a. manifestConstantDeclaration
    b. typeDeclaration
    c. collectionDeclaration
    d. converterDeclaration
    e. procedureDeclaration
    f. functionDeclaration
    g. moduleDeclaration
    h. monitorDeclaration

Each separateDeclaration can be a manifest constant declaration, a type declaration, a collection declaration, a converter declaration, a procedure or function declared as **external** in another compilation, or a module or monitor declared as **external** in another compilation.

Separately compiled procedures, functions, modules and monitors can be linked to form a complete program. Variables cannot be separately compiled and are not linked across compilations. Consistency of constants, types and collections is not automatically checked across compilations. Consistency of the type and number of formal parameters and function results between the external declaration and the separate compilation of separately compiled procedures and functions is not automatically checked.

Separately compiled modules and monitors will be initialized at the point of the corresponding **external** declaration. Note that since execution of a program consists of initializing the main module (see "Programs"), only those modules and monitors which are declared in the main module or a module nested within it will be initialized.

LINKING OF COMPILATIONS

A complete program will typically consist of a main module compilation linked together with the separate compilations of any procedures, functions, modules and monitors declared as **external** in it. The compilations must be linked such that the entry point of the program is the beginning of the main module compilation. (Under many systems, this means simply that the main module compilation must be the first in the list of object modules to be linked together.)

COLLECTED SYNTAX OF CONCURRENT EUCLID

The syntax of SE is given first. Throughout the following, {item} means zero or more of the item, and [item] means the item is optional.

The following abbreviations are used:

> id for identifier
> expn for expression
> typeDefn for typeDefinition

Semicolons are not required, but they may optionally appear following statements, declarations and import, export and **checked** clauses.

A *program* is:

> moduleDeclaration

A *moduleDeclaration* is:

> **var** id :
> > **module**
> > > [**imports** ([**var**] id {, [**var**] id})]
> > > [**exports** (id {, id})]
> > > [[**not**] **checked**]
> > > {declarationInModule}
> > > [**initially**
> > > > procedureBody]
> > **end** module

A *declarationInModule* is one of the following:

    a.   constantDeclaration
    b.   variableDeclaration
    c.   typeDeclaration
    d.   variableBinding
    e.   moduleDeclaration
    f.   collectionDeclaration
    g.   procedureDeclaration

    h. functionDeclaration
    i.   converterDeclaration
    j.   **assert** [(expn)]

A *constantDeclaration* is one of:

    a. [**pervasive**] **const** id := manifestExpn
    b. [**pervasive**] **const** id : typeDefn := expn
    c. [**pervasive**] **const** id : typeDefn :=
              (manifestExpn {, manifestExpn})
    d. [**pervasive**] **const** id := stringLiteral

A *manifestExpn* is:

    expn

A *variableDeclaration* is:

    [**register**] **var** id [(**at** manifestExpn)] : typeDefn
        [:= expn]

A *typeDeclaration* is:

    [**pervasive**] **type** id = typeBody

The *typeBody* is one of:

    a. typeDefn
    b. **forward**

A *typeDefn* is one of the following:

    a. standardType
    b. manifestConstant .. manifestExpn
    c. [**packed**] **array** indexType **of** typeDefn
    d. **set of** baseType
    e. [**packed**] recordType
    f.   pointerType
    g. namedType

A *standardType* is one of:

    a. SignedInt
    b. UnsignedInt
    c. LongInt
    d. ShortInt
    e. Boolean
    f.   Char
    g. StorageUnit
    h. AddressType

A *manifestConstant* is one of:

      a.   [-] literalConstant
      b.   [-] [moduleId] . manifestConstantId

A *manifestConstantId* is:

      id

An *indexType* is one of:

      a.   Char
      b.   manifestConstant .. manifestExpn
      c.   namedType

A *baseType* is one of:

      a.   0 .. manifestExpn
      b.   namedType

A *recordType* is:

      record
          var id : typeDefn
          {**var** id : typeDefn}
      end record

A *pointerType* is:

      ^ collectionId

A *collectionId* is:

      id

A *namedType* is:

      [moduleId .] typeId

A *moduleId* is:

      id

A *typeId* is:

      id

A *variableBinding* is one of:

      a.   **bind** [**register**] [**var**] id **to** variable
      b.   **bind** ([**register**] [**var**] id **to** variable
             {, [**register**] [**var**] id **to** variable})

A *collectionDeclaration* is:

> **var** id : **collection** of typeDefn

A *procedureDeclaration* is:

> **procedure** id [([**var**] id : parameterType
>   {, [**var**] id : parameterType})] =
> procedureBody

A *functionDeclaration* is:

> **function** id [(id : parameterType
>   {, id : parameterType})]
>   **returns** id : resultType =
> procedureBody

A *parameterType* is one of:

> a.   typeDefn
> b.   [**packed**] **array** manifestConstant **.. parameter** of
>   typeDefn
> c.   **universal**

A *resultType* is one of:

> a.   standardType
> b.   manifestConstant **..** manifestExpn
> c.   **set of** baseType
> d.   pointerType
> e.   namedType

A *procedureBody* is:

> [**imports** ([**var**] id {, [**var**] id})]
> **begin**
>   [[**not**] **checked**]
>   {declarationInRoutine}
>   {statement}
> **end** [id]

A *declarationInRoutine* is one of:

> a.   constantDeclaration
> b.   variableDeclaration
> c.   typeDeclaration
> d.   variableBinding
> e.   collectionDeclaration
> f.   converterDeclaration
> g.   **assert** [(expn)]

A *converterDeclaration* is:

converter id (typeId) **returns** typeId

A *statement* is one of:

a. variable := expn
b. [moduleId.] procedureId [(expn {, expn})]
c. **assert** [(expn)]
d. **return** [(expn)]
e. **if** expn **then**
        {statement}
    {**elseif** expn **then**
        {statement}}
    [**else**
        {statement}]
    end if
f. **loop**
        {statement}
    end loop
g. **exit** [**when** expn]
h. **case** expn **of**
        manifestExpn {, manifestExpn} = >
            {statement}
            end manifestExpn
        {manifestExpn {, manifestExpn} = >
            {statement}
            end manifestExpn}
    [**otherwise** = >
        {statement}]
    end case
i. **begin**
        {declarationInRoutine}
        {statement}
    end
j. collectionId . New (variable)
k. collectionId . Free (variable)

A *procedureId* is:

id

A *variable* is:

[moduleId .] id {componentSelector}

A *componentSelector* is one of:

    a.   (expn)
    b.   . id
    c.   . size
    d.   . address

An *expn* is one of the following:

    a.   variable
    b.   literalConstant
    c.   setTypeId (elementList)
    d.   collectionId . nil
    e.   [moduleId .] functionId [(expn {, expn})]
    f.   [moduleId .] converterId (expn)
    g.   (expn)
    h.   - expn
    i.   expn arithmeticOperator expn
    j.   expn comparisonOperator expn
    k.   **not** expn
    l.   expn booleanOperator expn
    m.   expn setOperator expn

A *setType* is:

    id

A *elementList* is one of:

    a.   [expn {, expn}]
    b.   **all**

A *functionId* is one of:

    a.   id
    b.   Chr
    c.   Ord
    d.   Long

A *converterId* is:

    id

An *arithmeticOperator* is one of:

    a.   +
    b.   -
    c.   *
    d.   **div**
    e.   **mod**

A *comparison Operator* is one of:

    a.   <
    b.   >
    c.   =
    d.   < =
    e.   > =
    f.   **not** =

A *boolean Operator* is one of:

    a.   **and**
    b.   **or**
    c.   ->

A *set Operator* is one of:

    a.   +
    b.   -
    c.   *
    d.   < =
    e.   > =
    f.   **in**
    g.   **not in**

Note: The order of precedence is among the following classes of operators (most binding first):

    1.   unary -
    2.   *, **div, mod**
    3.   +, -
    4.   <, >, =, < =, > =, **not** =, **in, not in**
    5.   **not**
    6.   **and**
    7.   **or**
    8.   ->

An *include Statement* is:

    include stringLiteral

Note: Include statements can appear anywhere in a program.

The following changes and additions are made to form CE:

A *module Declaration* is:

    var id :
        module
            [**imports** ([**var**] id {, [**var**] id})]
            [**exports** (id {, id})]

>   [[**not**] **checked**]
>   {declarationInModule}
>   [**initially**
>   procedureBody]
>   {**process** id [(memoryRequirement)]
>   procedureBody}
>   **end module**

A *memoryRequirement* is:

> manifestExpn

A *declarationInModule* is one of the following:

> a.   constantDeclaration
> b.   variableDeclaration
> c.   typeDeclaration
> d.   variableBinding
> e.   moduleDeclaration
> f.   monitorDeclaration
> g.   collectionDeclaration
> h.   procedureDeclaration
> i.   functionDeclaration
> j.   converterDeclaration
> k.   **assert** [(expn)]

A *monitorDeclaration* is:

> **var** id :
>   **monitor**
>   [**imports** ([**var**] id {, [**var**] id})]
>   [**exports** (id {, id})]
>   [[**not**] **checked**]
>   {declarationInMonitor}
>   [**initially**
>   procedureBody]
>   **end monitor**

A *declarationInMonitor* is one of the following:

> a.   constantDeclaration
> b.   variableDeclaration
> c.   typeDeclaration
> d.   variableBinding
> e.   conditionDeclaration
> f.   collectionDeclaration
> g.   procedureDeclaration
> h.   functionDeclaration

    i.   converterDeclaration
    j.   **assert** [(expn)]

A *conditionDeclaration* is one of:

    a.   **var** id : **[priority] condition**
    b.   **var** id : **array** indexType **of [priority] condition**

A *statement* is one of:

    a.   variable := expn
    b.   [moduleId.] procedureId [(expn {, expn})]
    c.   **assert** [(expn)]
    d.   **return** [(expn)]
    e.   **if** expn **then**
          {statement}
       {**elseif** expn **then**
          {statement}}
       [**else**
          {statement}]
       end if
    f.   **loop**
          {statement}
       end loop
    g.   **exit** [**when** expn]
    h.   **case** expn **of**
          manifestExpn {, manifestExpn} =>
             {statement}
             end manifestExpn
          {manifestExpn {, manifestExpn} =>
             {statement}
             end manifestExpn}
          [**otherwise** =>
             {statement}]
       end case
    i.   **begin**
          {declarationInRoutine}
          {statement}
       end
    j.   collectionId . New (variable)
    k.   collectionId . Free (variable)
    l.   **wait** (conditionVar [, priorityValue])
    m.   **signal** (conditionVar)
    n.   **busy** (time)

A *moduleId* is:

moduleOrMonitorId

A *moduleOrMonitorId* is:

> id

A *conditionVar* is:

> conditionId [(expn)]

A *conditionId* is:

> id

A *priorityValue* is:

> expn

A *time* is:

> expn

A *functionId* is one of:

> a. id
> b. Chr
> c. Ord
> d. Long
> e. empty

The following extensions allow separate compilation of procedures, functions, modules and monitors:

An *externalProcedureDeclaration* is:

> **procedure** id [([**var**] id : parameterType
> {, [**var**] id : parameterType})] =
> **external**

An *externalFunctionDeclaration* is:

> **function** id [(id : parameterType
> {, id : parameterType})]
> **returns** id : resultType =
> **external**

An *externalModuleDeclaration* is:

> **var** id :
> **external module**
> [**imports** ([**var**] id {, [**var**] id})]
> [**exports** (id {, id})]
> {declarationInExternalModule}
> **end module**

A *declarationInExternalModule* is one of:

    a. manifestConstantDeclaration
    b. typeDeclaration
    c. collectionDeclaration
    d. converterDeclaration
    e. externalProcedureDeclaration
    f. externalFunctionDeclaration

An *externalMonitorDeclaration* is:

    var id :
        external monitor
            [**imports** ([**var**] id {, [**var**] id})]
            [**exports** (id {, id})]
            {declarationInExternalMonitor}
        end monitor

A *declarationInExternalMonitor* is one of:

    a. manifestConstantDeclaration
    b. typeDeclaration
    c. collectionDeclaration
    d. converterDeclaration
    e. externalProcedureDeclaration
    f. externalFunctionDeclaration

Note: An external declaration can appear in place of the real declaration anywhere in a program.

A *manifestConstantDeclaration* is:

    [**pervasive**] **const** id := manifestExpn

A *separateCompilation* is:

    {separateDeclaration}

Each *separateDeclaration* is one of the following:

    a. manifestConstantDeclaration
    b. typeDeclaration
    c. collectionDeclaration
    d. converterDeclaration
    e. procedureDeclaration
    f. functionDeclaration
    g. moduleDeclaration
    h. monitorDeclaration

KEYWORDS AND PREDEFINED IDENTIFIERS

The following are reserved words of Euclid. These must not be used as identifiers in SE and CE programs. Those which are not in the SE subset are marked with an *.

*abstraction	*aligned	all	and
*any	array	assert	at
begin	bind	*bits	*bound
case	*checkable	checked	*code
collection	const	converter	*counted
*decreasing	*default	*dependent	div
else	elseif	end	exit
exports	*finally	*for	forward
*from	function	if	imports
in	include	initially	*inline
*invariant	loop	machine	mod
not	of	or	otherwise
packed	parameter	pervasive	*post
*pre	procedure	*readonly	record
return	returns	set	then
*thus	to	type	*unknown
var	when	*with	*xor

The following are additional reserved words of SE and CE. These also must not be used as identifiers in SE and CE programs.

busy	condition	empty	monitor
priority	process	register	signal
universal	wait		

The following are predefined identifiers of Euclid. In general, these are pervasive and must not be redeclared in SE and CE programs. Those which are not in the SE subset are marked with an *.

*Abs	address	AddressType	*alignment
*BaseType	Boolean	Char	Chr
*ComponentType	false	*first	Free
*Index	*IndexType	*Integer	*itsTag
*ItsType	*last	*Max	*Min
New	nil	*ObjectType	*Odd
Ord	*Pred	*refCount	SignedInt
size	*sizeInBits	StorageUnit	*String
*StringIndex	*stringMaxLength	*Succ	
*SystemZone	true	UnsignedInt	

The following are additional predefined identifiers of SE and CE. These also must not be redeclared in SE and CE programs.

Long	LongInt	ShortInt

INPUT/OUTPUT IN CONCURRENT EUCLID

This paper presents the standard input/output package for SE and CE. The user can access the I/O facility by including in his program the stub input/output module which corresponds to the level of I/O which his program requires. In this way, the user's compiled and linked program will include code only for the I/O facilities required.

The package provides four levels of sophistication, which are called "IO/1" through "IO/4". Each level includes all the facilities of the previous levels plus certain new features. The levels are as follows:

IO/1: Terminal (standard) input and output; Formatted text input/output of integers, characters and strings (Get and Put).

IO/2: Program argument sequential files; Open and close on argument files; Formatted text input/output of integers, characters and strings to files (FGet and FPut); Internal representation input/output of integers, characters and strings to files (Read and Write); End of file detection (EndFile).

IO/3: Temporary and non-argument sequential files (Assign, Deassign, Delete); Program arguments (FetchArg); Program error exit (SysExit).

IO/4: Record, array and storage input/output (Read and Write); Random access files (Tell and Seek); Error detection (Error).

The procedures and functions of the input/output system are all part of the module "IO" and must be referenced using "IO.". The types and constants which form the interface to the module are global. The user can access the level n facilities of the input/output module by including the statement

include '/usr/lib/coneuc/IOn'

as the first declaration in his main module.

We now describe the input/output facilities in detail.

IO/1: Terminal Formatted Text I/O

pervasive const newLine := $$N
pervasive const endOfFile := $$E
pervasive const maxStringLength :=
{ Implementation defined; >= 128 }
Strings read and written by the input/output routines may be up to maxStringLength characters in length.

IO.PutChar (c: Char)
Prints the character c on the terminal.

IO.PutInt (i: SignedInt, w: SignedInt)
> Prints the integer i on the terminal, right justified in a field of w characters. Leading blanks are supplied to fill the field. If w is an insufficient width, the value is printed in the minimum possible width with no leading blanks. In particular, if w is 1 then the exact number of characters needed is used. The specified width must be greater than zero and less than maxStringLength.

IO.PutLong (i: LongInt, w: SignedInt)
> Same as IO.PutInt for long integers.

IO.PutString (s: **packed array** 1..**parameter of** Char)
> Prints the string s on the terminal. The string must be terminated by an endOfFile character ('$E'), which is not output. It can contain embedded newLines ('$N') if desired. (Note: An endOfFile character ($$E) can be output using PutChar.)

IO.GetChar (**var** c:Char)
> Gets the next input character from the terminal. End of file is indicated by a return of endOfFile ($$E).

IO.GetInt (**var** i: SignedInt)
> Gets an integer from the terminal. The input must consist of any number of optional blanks, tabs and newlines, followed by an optional minus sign, followed by any number of decimal digits.

IO.GetLong (**var** i: LongInt)
> Same as IO.GetInt for long integers.

IO.GetString (**var** s: **packed array** 1..**parameter** of Char)
> Gets a line of character input from the terminal. The returned string may be up to maxStringLength characters in length. The string returned is ended with the newLine character ('$N') followed by an endOfFile character ('$E') if it is a complete line, and by the endOfFile character only if it is a partial line (i.e., if the input line exceeds maxStringLength characters in length). End of file is indicated by returning a string containing endOfFile ('$E') as the first character.

IO/2: Sequential Argument File I/O

pervasive const stdInput := -2
pervasive const stdOutput := -1
pervasive const stdError := 0
pervasive const maxArgs := { Implementation defined; $>= 9$ }
pervasive const maxFiles :=

{ Implementation defined; $> = $ maxArgs$+5$ }

type File $= $ stdInput..maxFiles

Concurrent Euclid input/output refers to files using a file number. Certain file numbers are preassigned as follows: -2 refers to the terminal input; -1 is the terminal output; 0 is the standard diagnostic output. The file numbers 1..maxArgs refer to the program arguments. The remaining file numbers (maxArgs$+1$..maxFiles) can be dynamically assigned to files using the "IO.Assign" operation; see "IO/3".

pervasive const inFile $:= 0$
pervasive const outFile $:= 1$
pervasive const inOutFile $:= 2$
type FileMode $= $ inFile..inOutFile

Files can be opened for input, output, or input/output using modes inFile, outFile and inOutFile respectively. (Note: The input/output mode is not available under Unix V6.)

IO.Open (f: File, m: FileMode)
IO.Close (f: File)

With the exception of terminal input/output and the standard diagnostic output, files must be opened before they are used and closed before the program returns. Open opens an existing file for the operations specified by the mode. If the opened file does not exist, it is created. The file number specified must be a preassigned file number or a file number returned from a call to "IO.Assign"; see "IO/3".

IO.FPutChar (f: File, c: Char)
IO.FPutInt (f: File, i: SignedInt, w: SignedInt)
IO.FPutLong (f: File, i: LongInt, w: SignedInt)
IO.FPutString (f: File, s: **packed array** 1..**parameter of** Char)
IO.FGetChar (f: File, **var** c: Char)
IO.FGetInt (f: File, **var** i: SignedInt)
IO.FGetLong (f: File, **var** i: LongInt)
IO.FGetString (f: File, **var** s: **packed array** 1..**parameter of** Char)

These operations are identical to the terminal input/output operations of IO/1 except that the put or get is done on the specified file.

IO.WriteChar (f: File, c: Char)

Writes the internal representation of character c to the specified file.

IO.WriteInt (f: File, i: SignedInt)

Writes the internal representation of integer i to the specified file.

IO.WriteLong (f: File, i: LongInt)
> Writes the internal representation of long integer i to the specified file.

IO.WriteString (f:File, s: **packed array** 1..**parameter of** Char)
> Writes the internal representations of the characters in the string s to the specified file.

IO.ReadChar (f: File, **var** c: Char)
> Reads a character in internal representation from the specified file into c.

IO.ReadInt (f: File, **var** i: SignedInt)
> Reads an integer in internal representation from the specified file into i.

IO.ReadLong (f: File, **var** i: LongInt)
> Reads a long integer in internal representation from the specified file into i.

IO.ReadString (f: File, **var** s: **packed array** 1..**parameter** of Char)
> Reads a string of characters terminated by a newLine character ('$N') in internal representation from the specified file into s. The returned string may be up to maxStringLength characters in length. The string returned is ended with the newLine character ('$N') followed by an endOfFile character ('$E') if it is a complete line, and by the endOfFile character only if it is a partial line (i.e., if the input line exceeds maxStringLength characters in length). End of file is indicated by returning a string containing endOfFile ('$E') as the first character.

IO.EndFile (f: File)
> A function which returns true if the last operation on the specified input file encountered end of file and false otherwise.

IO/3: Temporary and Non-argument Files

pervasive const maxArgLength :=
> { Implementation defined; $> = 32$ }
> File names and arguments to a program may be up to maxArgLength characters in length.

IO.Assign (**var** f: File, s: **packed array** 1..**parameter** of Char)
> A file number is assigned to the file name supplied in s. The file name is given as a string terminated by the endOfFile character ('$E'), which is not part of the name. Before the file can be used it must be opened using "IO.Open".

IO.Deassign (f: File)
> The specified file number is freed for assignment to another file name. An open file cannot be deassigned.

IO.Delete (f: File)
> The specified file is destroyed. An open file cannot be deleted. Note that a program can have temporary files using "IO.Assign" and "IO.Delete".

IO.FetchArg (n: 1..maxArgs, **var** s: **packed array** 1..**parameter of** Char)
> The program argument specified by "n" is returned in string s. The returned string is terminated by the endOfFile character ('$E') and may be up to maxArgLength characters in length.

IO.SysExit (n: SignedInt)
> Terminate program execution with the specified return code. (CE programs return 0 by default.)

IO/4: Structure Input/Output and Random Access Files

IO.Write (f: File, u: **universal**, n: SignedInt)
> The number of StorageUnits specified by "n" are written to the file from u. Write can be used to write out whole arrays and records using a call of the form "IO.Write (f, v, v.size)". The value of n must be positive or zero.

IO.Read (f: File, **var** u: **universal**, n: SignedInt)
> The number of StorageUnits specified by "n" are read from the file into u. Read can be used to read in whole arrays and records using a call of the form "IO.Read (f, v, v.size)". The value of n must be positive or zero.

type FileIndex = LongInt
IO.Tell (f: File, **var** x: FileIndex)
IO.Seek (f: File, x: FileIndex)
> These operations provide random access input/output by allowing the program to sense a file position, represented as a long integer, and reset the file to a remembered position. Tell returns the current position of the specified file. Seek sets the current position of the specified file to the position specified by the value of x. The representation of file indices is implementation-dependent. (Note: "IO.Tell" and "IO.Seek" are not supported under Unix V6.)

IO.Error (f: File)
> A function which returns true if the last operation on the specified file encountered an error and false otherwise.

Interfacing to Unix

The input/output package is based on standard Unix input/output and is designed to be interfaced to Unix with a minimum of overhead. The Unix implementation is written in C and uses only facilities of the C "stdio" package. This implementation can be compiled unchanged under both V6 and V7 Unix.

PDP-11 IMPLEMENTATION NOTES

This section gives details of the implementation of CE for the PDP-11 under Unix and provides information necessary for interfacing with CE programs.

DATA REPRESENTATION

The following gives the storage representations of the various CE data types used by the PDP-11 implementation.

Type	Representation
SignedInt and subranges contained in -32768..32767	16-bit signed
UnsignedInt and subranges contained in 0..65535 but outside -32768..32767	16-bit unsigned
LongInt and subranges outside the above	32-bit signed 16-bit aligned; high order word has the lower address
ShortInt and packed subranges in 0..255	8-bit unsigned
Boolean	8-bit unsigned true = 1, false = 0
Char	8-bit unsigned
StorageUnit	8-bit unsigned
AddressType, pointers and binds	16-bit unsigned
sets of 0..7	8-bit unsigned; element 0 is low order bit, element 7 is high order bit
sets of 0..15	16-bit unsigned; element 0 is low order bit, element 15 is high order bit

REGISTER USAGE

The following register assignments are used by the PDP-11 implementation.

Register	Use
R0, R1	function results, scratch
R2, R3	scratch
R4 and binds	line number, register variables
R5	register variables and binds

Since the CE implementation uses the stack pointer register (SP) to address local variables in procedures and functions, there is no local base register.

Function results whose data representation is a byte or word are returned in R0. Doubleword results are returned in R0 and R1, with the high order word in R0.

In order to attain highly efficient code for non-scalar assignments, subscripting and LongInt arithmetic, the CE compiler uses four scratch registers rather than the two used by the C compiler. In particular, CE uses R2 and R3 for scratch and hence does not save and restore them at procedure and function entry and exit. Since the PDP-11 C compiler uses R2 and R3 for **register** variables, C routines which call CE procedures and functions can use at most one **register** variable. There is no such restriction on C routines called from CE programs.

Register R5 (and R4 when line numbering is turned off, see below) are used for user variables and binds which are explicitly declared **register**.

When run-time line numbering is turned on (which is the default), the CE compiler generates code to maintain the source file and line number in the line number register (R4) during execution. This aids in debugging since the "cedb" program can obtain the source file name and line number from the core dump following a run-time program failure (e.g., assertion failure, subscript or case tag out of range, etc.).

The contents of the line number register is interpreted as a 5 digit unsigned decimal number, the first two digits of which give the source **include** file number and the last three of which give the source line number within file. Source file numbers are assigned sequentially starting with 1 for the main source file. Source files longer than 999 lines are assigned a new file number for each 1000 lines of source.

Run time line numbering can be turned off using the "-l" compiler toggle.

CALLING CONVENTIONS

CE procedures and functions which are (a) declared **external**, (b) separately compiled, or (c) exported from a separately compiled module or monitor, are called using the C calling convention. A more efficient calling convention is used for calls between CE routines within a single compilation.

Unlike C routines, CE procedures and functions do not save and restore all of the caller's registers, but rather save and restore only those registers which they actually use. Note that since registers R0-R3 are considered scratch registers by the CE compiler, CE routines never save and restore R0-R3. This means that C routines which call CE routines can use at most one **register** variable. C routines which are called from CE may of course use as many **register** variables as they wish. Assembly routines called from CE can use R0-R3 as scratch and need not save and restore them. (Exception: the CE built-in routines are called using a special calling convention and must save and restore all registers which they use).

EXTERNAL NAMES

CE procedures and functions which are (a) declared "external", (b) separately compiled, or (c) exported from a separately compiled module or monitor, are assigned external names so that they may be linked with and/or called from other compilations and programs. On the PDP-11 under Unix, these names consist of the routine name preceded by an underscore character. Because of Unix linker restrictions, only the first seven characters of external names are significant and hence care must be taken to avoid conflilcts. The **initially** routine of an external module or monitor is given the name of the module/monitor.

PARAMETER PASSING

Like C, CE passes parameters on the PDP-11 stack. Unlike C, however, CE pushes parameters onto the stack in the order in which they appear in the call (C reverses this order). Hence C procedures and functions which are called from CE (and CE procedures and functions which are called from C) must declare their formal parameters in reversed order.

Value parameters as defined in the CE language specification are passed as values on the stack. Byte values are passed in the low order byte of a 16-bit word. Reference parameters are passed as 16-bit word addresses.

A parameter passed to an array formal parameter declared using the **parameter** keyword as its upper bound is passed with an extra unsigned word parameter following the array address. This extra parameter gives the number of elements in the array minus one. A parameter passed to a **universal** formal parameter is passed as an address only.

RUN-TIME CHECKING

When run-time checking is turned on (which is the default), the CE compiler will generate code to check **assert** statements, subscript ranges and case selector ranges during execution. It will not generate code to check ranges in assignments and overflow in expressions at run-time. The checking code uses an illegal instruction of the form "jsr r0,rN" to abort the program when a run-time check fails. The second register number in the instruction is an abort code indicating the reason for the abort. The following table gives the abort codes used by the PDP-11 implementation.

Aborting instruction	Reason for abort
jsr r0,r0	assertion failure
jsr r0,r1	subscript out of range
jsr r0,r2	case selector out of range
jsr r0,r3	function failed to return a value

The "cedb" utility will automatically determine the source file name, source line number and reason for abort from the core file produced by a run-time abort.

All run-time checking can be turned off using the "-k" compiler toggle.

CONCURRENT EUCLID IMPLEMENTATION NOTES

This section gives the details of the implementation of CE for the PDP-11, VAX, MC68000 and MC6809 and provides information necessary for interfacing with CE programs on these machines.

DATA REPRESENTATION

Sets and the predefined types (except AddressType) have the same representations across the four machines.

Type	*Representation*
ShortInt	8-bit unsigned
SignedInt	16-bit signed
UnsignedInt	16-bit unsigned
LongInt	32-bit signed
set of 0..n, n < 8	8-bit
set of 0..n, n > 7	16-bit
Boolean	8-bit (0=false, 1=true)
Char and subranges of Char	8-bit unsigned

AddressType and pointers have sizes the same as the target machine's address size (16 bits for the PDP-11 and MC6809 and 32 bits for the MC68000 and VAX). Register variables which are actually allocated to registers use the most efficient (fastest) representation for their range on the target machine (16-bit words are favored on the PDP-11, MC68000 and MC6809 and 32-bit longwords on the VAX). Subrange elements of packed arrays and records have the same representations across the four machines; if the subrange is within ShortInt, an unsigned byte is used; otherwise if it is within SignedInt, a 16-bit signed word is used; otherwise if it is within UnsignedInt, a 16-bit unsigned word is used; otherwise a 32-bit signed longword is used. Subrange elements of unpacked arrays and records have the same representations as packed subranges except that unsigned bytes are not used. The MC6809 treats all records and arrays as packed.

The following table summarizes the data representations of the various CE data types used by the implementations.

Type	PDP-11	VAX	MC68000	MC6809
ShortInt	8-bit unsigned	8-bit unsigned	8-bit unsigned	8-bit unsigned
SignedInt	16-bit signed	16-bit signed	16-bit signed	16-bit signed
UnsignedInt	16-bit unsigned	16-bit unsigned	16-bit unsigned	16-bit unsigned
LongInt	32-bit signed	32-bit signed	32-bit signed	32-bit signed
Boolean	8-bit	8-bit	8-bit	8-bit
Char and Char subranges	unsigned 8-bit	unsigned 8-bit	unsigned 8-bit	unsigned 8-bit
StorageUnit	8-bit	8-bit	8-bit	8-bit
AddressType, pointers and bind pointers	16-bit unsigned	32-bit signed	32-bit signed	16-bit unsigned
sets of 0..7	8-bit	8-bit	8-bit	8-bit
sets of 0..15	16-bit	16-bit	16-bit	16-bit
Packed subrange in 0..255	8-bit unsigned	8-bit unsigned	8-bit unsigned	8-bit unsigned
Unpacked subrange in 0..255	16-bit signed	16-bit signed	16-bit signed	8-bit unsigned
Packed/unpacked subrange in -32768..32767	16-bit signed	16-bit signed	16-bit signed	16-bit signed
Packed/unpacked subrange in 0..65535	16-bit unsigned	16-bit unsigned	16-bit unsigned	16-bit unsigned
Subrange not in any of above	32-bit signed	32-bit signed	32-bit signed	32-bit signed

Register variable in -32768..32767	16-bit signed	32-bit signed	16-bit signed	(Not Applicable)
Register variable in 0..65535	16-bit unsigned	32-bit signed	16-bit unsigned	(Not Applicable)
Register variable not in any of above	(Not Applicable)	32-bit signed	32-bit signed	(Not Applicable)

STORAGE LAYOUT

Data types have different storage layouts on different machines. The following table gives the byte offsets from the base address (x.address) of the sub-bytes and sub-words of the data representation types.

Data representation	*PDP-11*	*VAX*	*MC68000*	*MC6809*
16-bit word				
high order byte	1	1	0	0
low order byte	0	0	1	1
32-bit longword				
high order word	0	2	0	0
low order word	2	0	2	2
high order byte	1	3	0	0
2nd highest order byte	0	2	1	1
2nd lowest " "	3	1	2	2
low order byte	2	0	3	3

Set elements are represented using one bit per element. Element 0 is represented by the low order bit, element 1 by the next bit and so on.

Alignment of the data representations also varies among the machines. The following table gives the alignments of the various representations.

Data representation	*PDP-11*	*VAX*	*MC6800*	*MC6809*
byte	byte	byte	byte	byte
16-bit word	16-bit	16-bit	16-bit	byte
32-bit longword	16-bit	32-bit	16-bit	byte

REGISTER USAGE

The following register assignments are used by the various implementations.

Register use	PDP-11	VAX	MC68000	MC6809
Data scratch	R0-R3	R0-R5	D0-D2	A,B,D,X
Address scratch	"	"	A1-A3	X,Y
Function result	R0,R1	R0	D0	D,X
Line number	R4	R6	D3	U
Register variables	R5,R4[*]	R7-R11, R6[*]	D4-D7, D3[*]	(not applicable)
Register binds	"	"	A4-A5	(not applicable)

[*] - when line numbering is turned off

INDEX